INORGANIC
SYNTHESES
Volume XIV

Editors

AARON WOLD

*Professor of Engineering
and Chemistry
Brown University, Providence, R.I.*

JOHN K. RUFF

*Associate Professor of Chemistry
University of Georgia
Athens, Ga.*

INORGANIC SYNTHESES

Volume XIV

McGRAW-HILL BOOK COMPANY

*New York   St. Louis   San Francisco   Düsseldorf
Johannesburg   Kuala Lumpur   London   Mexico
Montreal   New Delhi   Panama   Rio de Janeiro
Singapore   Sydney   Toronto*

To RONALD NYHOLM and DAVID WADSLEY

CONTENTS

PREFACE

This volume of INORGANIC SYNTHESES has a large section (Chapter Four) devoted to solid-state syntheses. Many of the previous volumes have contained solid-state preparations, and it is our hope that this volume will encourage chemists working in this area to contribute manuscripts suitable for the series. An important aspect of Chapter Four is the inclusion of several syntheses which deal with the preparation and characterization of single crystals. The preparation of well-characterized single crystals, suitable for physical measurements, is essential if the results of such measurements are to be of interest either from an industrial or academic point of view. In addition to chemical analyses and spectral data, other physical measurements may be used to identify the nature of the products formed from solid-state reactions. These include microscopic identification, x-ray analyses, density determination, and magnetic, electrical, and optical measurements. These are chosen in sufficient combination to provide ample evidence that the product is single-phase, stoichiometric, and homogeneous.

The arrangement of the syntheses in the first part of Volume XIV is divided into three chapters, namely, Phosphorus Compounds, Non-Transition-Metal Compounds, and Transition-Metal Compounds. The reader is advised to seek particular compounds in the subject or formula indexes. The indexes at the end of this volume are cumulative from Volume XI through Volume XIV.

Inorganic Syntheses, Inc. is a nonprofit organization whose goals are to further interest in synthetic inorganic chemistry. Since the preparations are detailed and checked, they can probably be carried out even by uninitiated workers in a new field. The generous cooperation of inorganic chemists is necessary for the continued success of this series.

We should like to thank many of the members of Inorganic Syntheses, Inc. for their generous help in making this volume possible. In particular, we thank Professors W. C. Fernelius, S. Kirschner, T. Moeller, S. Y. Tyree, Dr. G. Parshall, and Dr. W. H. Powell for the many hours they have spent helping us with the perplexing problems concerning the checking difficulties. We should like to thank Dr. D. B. Rogers and his associates at E. I. du Pont de Nemours & Company, and our own students as well, who have helped us prepare this volume. Finally, we would like to thank Janet W. Cherry for her contributions in preparing the manuscript.

A. Wold
J. Ruff

NOTICE TO CONTRIBUTORS

The INORGANIC SYNTHESES series is published to provide all users of inorganic substances with detailed and foolproof procedures for the preparation of important and timely compounds. Thus the series is the concern of the entire scientific community. Inorganic Syntheses, Inc. hopes that all chemists will share in the responsibility of producing INORGANIC SYNTHESES by offering their advice and assistance both in the formulation and laboratory evaluation of outstanding syntheses. Help of this type will be invaluable in achieving excellence and pertinence to current scientific interests.

There is no rigid definition of what constitutes a suitable synthesis. The major criterion by which syntheses are judged is the potential value to the scientific community. An ideal synthesis is one which presents a new or revised experimental procedure applicable to a variety of related compounds, at least one of which is critically important in current research. However, syntheses of individual compounds that are of interest or importance are also acceptable.

Inorganic Syntheses, Inc. lists the following criteria of content for submitted manuscripts. Style should conform with that of previous volumes of INORGANIC SYNTHESES. The *Introduction* should include a concise and critical summary of the available procedures for synthesis of the product in question. It should also include an estimate of the time required for the synthesis, and indication of the importance and utility of the product, and an admonition if any potential hazards are associated with the procedure. The *Procedure* should present detailed and unambiguous laboratory directions and be written so that it anticipates possible mistakes and misunderstandings on the part of the person who attempts to duplicate the procedure. Any unusual equipment or procedure should be clearly described. Line drawings should be included when they can be helpful. All safety measures should be clearly stated. *Sources of unusual starting materials must be given*, and, if possible, minimal standards of purity of reagents and solvents should be stated. The scale should be reasonable for normal laboratory operation, and any problems involved in scaling the procedure either up or down should be discussed. The criteria for judging the purity of the final product should be clearly delineated. The section on *Properties* should list and discuss those physical and chemical characteristics that are relevant to judging the purity of the product and to permitting its handling and use in an intelligent manner. Under *References*, all pertinent literature citations should be listed in order.

Inorganic Syntheses, Inc. determines whether submitted syntheses meet the general specifications outlined above. Every synthesis must be satisfactorily reproduced in a different laboratory other than that from which it was submitted.

Each manuscript should be submitted in duplicate to the Editorial Secretary, Professor Stanley Kirschner, Department of Chemistry, Wayne State University, Detroit, Michigan 48202, U.S.A. The manuscript should be typewritten in English. Nomenclature should be consistent and should follow the recommendations presented in "The Definitive Rules for Nomenclature of Inorganic Chemistry," *J. Am. Chem. Soc.*, 82, 5523 (1960). Abbreviations should conform to those used in publications of the American Chemical Society, particularly *Inorganic Chemistry*.

INORGANIC
SYNTHESES

Volume XIV

Chapter One

PHOSPHORUS COMPOUNDS

1. PHOSPHINE

$$2AlP + 3H_2SO_4 \longrightarrow 2PH_3 + Al_2(SO_4)_3$$

Submitted by ROBERT C. MARRIOTT,* JEROME D. ODOM,*
and CURTIS T. SEARS, JR.* †
Checked by V. D. BIANCO‡ and S. DORONZO‡

Phosphine has been prepared by the action of either water[1] or hydrochloric acid[2] on calcium phosphide or zinc phosphide;[3] the reaction of hot basic solutions on elemental phosphorus;[4] the pyrolysis of phosphorus acid;[5] and the action of sulfuric acid on aluminum phosphide.[6] The last method is the most convenient for the laboratory preparation of phosphine.

The method illustrated below is useful for the production of phosphine in quantities up to 5 g. It can be used to introduce the phosphine in the gas phase directly into a subsequent reaction. It is also adaptable to the synthesis of PD_3 by reaction of AlP with D_2SO_4–D_2O.

*University of South Carolina, Columbia, S.C. 29208.
†Present address: Department of Chemistry, Georgia State University, Atlanta, Ga. 30303.
‡Istituto di Chimica Generale ed Inorganica, Universita di Bari, Bari, Italy.

■ *Caution.   Phosphine is an extremely toxic gas. Maximum safe concentration for constant exposure is 0.3 p.p.m. of vapor in air. The crude reaction product is spontaneously flammable in air*[7] *because of the presence of trace amounts of* P_2H_4. *The reaction and the cleaning of the reaction vessel should be carried out in an efficient fume hood.*

Procedure

A 500-ml., three-necked, round-bottomed flask is equipped with a magnetic stirrer, a Claisen adapter in which are inserted two pressure-equalizing addition funnels, a nitrogen source, and an outlet to a series of traps (Fig. 1). Trap 1 is cooled to −78°C. (Dry Ice–acetone mixture) and is attached to a drying tube filled with potassium hydroxide. Traps 2 and 3, equipped with stopcocks, are cooled to −196°C. (liquid nitrogen*). All connections are made of

---

*If it is desired to have the phosphine as it is generated in the gas phase introduced into a subsequent reaction mixture, then replace traps 2 and 3 with the appropriate reaction vessels. The phosphine may contain a small trace of P_2H_4 at this point.

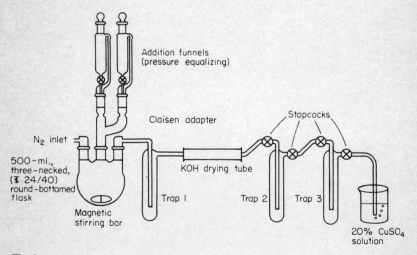

Addition funnels
(pressure equalizing)

Claisen adapter

Stopcocks

N_2 inlet

500–ml.,
three–necked,
(⦙ 24/40)
round–bottomed
flask

KOH drying tube

Trap 1   Trap 2   Trap 3

Magnetic
stirring bar

20% $CuSO_4$
solution

Fig. 1

Tygon tubing. The tubing exiting from trap 3 is immersed in a 20% solution of $CuSO_4$ to remove any traces of phosphine which are not condensed. The round-bottomed flask is charged with 1.18 g. (0.02 mole) of aluminum phosphide,* 10 ml. of water is placed in one funnel, and 50 ml. of 6 M sulfuric acid is placed in the other funnel. The entire system is flushed with nitrogen for 10 minutes. The nitrogen flow is reduced to a very slow flow rate † (one bubble per second from the $CuSO_4$ solution). Enough water (5-7 ml.) to cover the aluminum phosphide is added quickly to the reaction flask without stirring. Evolution of the phosphine begins immediately. After the initially vigorous reaction has subsided (approximately 5 minutes), stirring is started, and 5 ml. of 6 M sulfuric acid is added dropwise over a period of 15 minutes. The rate of addition of sulfuric acid is increased so that the remainder is added during a period of 10 minutes. After the addition is complete, the reaction mixture is stirred an additional 20 minutes to ensure complete reaction.

The nitrogen flow is stopped, and the two liquid-nitrogen traps are isolated. The phosphine is purified by evacuating the traps (still at $-196°C$.) and distilling into a vacuum line through a $-131°C$. trap (n-C_5H_{12} slush) into a $-196°C$. trap. Typical yields by this method are 75-77%.

Properties

Phosphine prepared by this method exhibits a vapor pressure of 170 ± 1 mm. at $-111.6°C$. (CS_2 slush). The literature value is 171 mm.[8] Infrared[9] and mass spectral[10] data have been reported. The infrared spectrum shows ν_{PH} at 2327 cm.$^{-1}$ and also peaks at 1121 and 900 cm.$^{-1}$. In the gas phase, all these bands show complex fine structure.

---

*Ventron Corporation, P.O. Box 159, Beverly, Mass. 01915.

†If the flow rate of nitrogen is too fast, a significant amount of phosphine is swept through the liquid-nitrogen traps.

References

1. D. T. Hurd, "Introduction to the Chemistry of Hydrides," p. 127, John Wiley & Sons, Inc., New York, 1952.
2. R. Paris and P. Tarde, *Compt. Rend.*, **233**, 242 (1946).
3. J. W. Elmore, *J. Assoc. Off. Agr. Chem.*, **26**, 559 (1943).
4. J. Datta, *J. Indian Chem. Soc.*, **29**, 965 (1952).
5. S. D. Gokhale and W. L. Jolly, *Inorganic Syntheses*, **9**, 56 (1967).
6. W. E. White and A. H. Bushey, *J. Am. Chem. Soc.*, **66**, 1666 (1944).
7. "Threshold Limit Values of Airborne Contaminants," adopted by ACGIH for 1969, American Conference of Governmental Industrial Hygienists, 1014 Broadway, Cincinnati, Ohio 45202.
8. S. R. Gunn and L. G. Green, *J. Phys. Chem.*, **65**, 779 (1961).
9. D. A. Tierney, D. W. Lewis, and D. Berg, *J. Inorg. Nucl. Chem.*, **24**, 1165 (1962).
10. Y. Wada and R. W. Kiser, *Inorg. Chem.*, **3**, 175 (1964).

2. *tert*-BUTYLDICHLOROPHOSPHINE AND DI-*tert*-BUTYLCHLOROPHOSPHINE

$$(CH_3)_3 CMgCl + PCl_3 \longrightarrow (CH_3)_3 CPCl_2 + MgCl_2$$
$$2(CH_3)_3 CMgCl + PCl_3 \longrightarrow [(CH_3)_3 C]_2 PCl + 2MgCl_2$$

Submitted by M. FILD,* O. STELZER,* and R. SCHMUTZLER*
Checked by G. O. DOAK†

The reaction of Grignard reagents with phosphorus trichloride usually proceeds all the way to the tertiary phosphine, $R_3 P$.[1] Stepwise alkylation is rarely observed. Examples of the latter include the formation of small amounts of the monochlorophosphines, $(n\text{-}C_8 H_{17})_2 PCl$[2] and $(C_6 H_{11})_2 PCl$,[3] in the reaction of the appropriate Grignard reagents with phosphorus trichloride. In

*Lehrstuhl B für Anorganische Chemie der Technischen Universität, Pockelsstrasse 4, D 33, Braunschweig, Germany.
†Department of Chemistry, North Carolina State University, Raleigh, N.C. 27607.

these cases, there is evidence that introduction of the third hydrocarbon group is difficult for steric reasons.

The nonformation of tertiary phosphines is particularly evident in the reaction of phosphorus trichloride with the Grignard reagents obtained from alkyl halides containing a branched primary, secondary, or tertiary alkyl group.[4,5,7,8] Thus, by using the appropriate molar ratio, PCl_3/RMgX, a convenient preparation of the chlorophosphines, $(CH_3)_3 CPCl_2$ and $[(CH_3)_3 C]_2 PCl$, is possible from phosphorus trichloride and *tert*-butylmagnesium chloride (*tert*-butylchloromagnesium).

Procedure

A. *tert*-BUTYLDICHLOROPHOSPHINE

1. Preparation of the Grignard Reagent

In a 1-l., three-necked flask, 100 ml. of absolute diethyl ether and 9.2 g. (0.1 mole) of freshly distilled *tert*-butyl chloride are added to 24.3 g. (1 g. atom) of magnesium turnings. The flask is equipped with a mechanical stirrer, reflux condenser, and a 250-ml. dropping funnel. The reaction is initiated by adding a few drops of bromine. The rest of the *tert*-butyl chloride (83.3 g., 0.9 mole), which has been dissolved in 200 ml. of ether, is added through the dropping funnel at such a rate that the reaction mixture remains at its boiling point. After the chemicals have been added, the mixture is maintained at reflux for an additional hour. The Grignard solution is then poured, under a nitrogen atmosphere, through a bent tapped adapter into a 500-ml. dropping funnel with a pressure-equalizing side arm. A small amount of glass wool is placed in the bottom of the dropping funnel in order to prevent clogging of the stopcock.

2. Preparation of *tert*-Butyldichlorophosphine

The reaction is conducted in an apparatus identical with that used in the preparation of the Grignard reagent. Phosphorus trichloride (137 g., 1 mole), dissolved in 600 ml. of ether, is placed in a 2-l., three-necked flask. The Grignard solution, contained in the 500-ml. dropping funnel, is added dropwise with stirring to the phosphorus trichloride over a period of 3 hours. A stream of nitrogen is slowly passed through a T tube on top of the reflux condenser, while a bath temperature between −20 and −10°C. is maintained during the addition period.

The temperature of the reaction mixture is allowed to rise to room temperature. The mixture is then refluxed for 1 hour. Solids are removed by filtration through a coarse sintered-glass, fritted funnel.* The residue is washed with two 100-ml. portions of ether. Ether and other volatile products are removed by distillation, first at atmospheric pressure, then *in vacuo* until a pressure of 40 mm. has been reached. The residual ether solution is added in portions to a 100-ml. flask assembled for distillation through a 12-in. Vigreux column. Distillation of the higher-boiling residue gives *tert*-butyldichlorophosphine as a colorless liquid of b.p. 142–145°C., which solidifies on standing. The yield† is 70–80 g. (44–50%), based on *tert*-BuMgCl.

B. DI-*tert*-BUTYLCHLOROPHOSPHINE

1. Preparation of the Grignard Reagent

The Grignard reagent is prepared in a 2-l., three-necked flask, fitted with a 500-ml. pressure-equalizing dropping funnel, a reflux condenser with a drying tube, and a mechanical stirrer. Magnesium turnings (24.3 g., 1 g. atom) and *tert*-butyl chloride (92.5 g., 1 mole) contained in a total 1300 ml. of dry ether are employed. After the initial addition of 400 ml. of ether, the magnesium

---

*See, e.g., D. F. Shriver, "The Manipulation of Air-sensitive Compounds," p. 147, Figs. 7, 9; McGraw-Hill Book Company, New York, 1969.

†The checker suggests a second distillation in order to obtain a very pure product.

turnings are etched with a trace of iodine. The reaction is then started by the addition of *tert*-butyl chloride. Subsequently, the ether solution of *tert*-butyl chloride is added gradually so that the ether solvent is kept at its boiling point (*ca.* 3 hours). To complete the reaction, the mixture is allowed to reflux for another hour.

The immediate use of the Grignard reagent thus obtained is imperative. If it is allowed to stand for any period of time, the yield in the subsequent reaction with phosphorus trichloride is reduced drastically. Furthermore, if formation of larger amounts of a white precipitate is observed at the end of the Grignard reaction, the reaction mixture should be discarded.

2. Preparation of Di-*tert*-butylchlorophosphine

This preparation is conducted in the same flask as that in which the Grignard reagent has been prepared. Phosphorus trichloride (34.4 g., 0.25 mole, in 50 ml. of ether) is added at room temperature with stirring to the Grignard reagent. A precipitate is formed readily. The reaction mixture is boiled and is allowed to reflux for 2 hours. Then the ether solution containing the product is separated from the precipitate by careful decanting into a 500-ml. dropping funnel in which some glass wool has been placed. This operation is conducted in a countercurrent of nitrogen. The decanted product is then added in portions through a 12-in. Vigreux column to a small flask set up for distillation at atmospheric pressure. The oily, yellow product thus remaining is distilled *in vacuo*. The product obtained is a colorless liquid of b.p. 70–72°C. (13 mm.); 100–102°C. (48 mm.). The yield ranges from 25 to 31.5 g. (55–70%).

Properties and Uses

Both chloro phosphines, *tert*-$C_4H_9PCl_2$ and (*tert*-C_4H_9)$_2$PCl, are very sensitive to air and moisture and must be handled in a

TABLE I   Physical Data for $tert$-$C_4H_9PCl_2$ and $(tert$-$C_4H_9)_2PCl$

Compound	m.p., °C.	b.p., °C.	δP, p.p.m.*†	δH, p.p.m.‡	$^3J_{P-H}$, Hz.
$tert$-$C_4H_9PCl_2$	49[4]	145-150 (760 mm)[4]	−198.6[11]	−1.07[11]	15.0[9]
	51.5-52.5[5]	140-145[9]			15.0[11]
	44-48[14]	139[14] (60/12 mm)[14]	−200[14]	−1.22[14]	15[14]
$(tert$-$C_4H_9)_2PCl$	2-3[8]	69-70 (10 mm)[4]	−145.0[11] §	−1.19[11]	12.0[8]
		70-72 (13 mm)[7]			12.2[9]
		48 (3 mm)[8]			12.0[11]
		68-69 (9 mm)[9]			

*Relative to an external H_3PO_4 standard.
†[1] H and [31] P spectra recorded in benzene solution.
‡Relative to internal $Si(CH_3)_4$.
§Neat liquid.

nitrogen atmosphere at all times. Distillation is the best method of purification for both compounds, but it has been reported that $tert$-$C_4H_9PCl_2$ may also be sublimed *in vacuo* (unspecified pressure) at 25°C.[5]

Physical data, such as melting and boiling points (m.p., b.p.), and [1] H and [31] P n.m.r. data, reported by various investigators, are listed in Table I.

The $tert$-butylchlorophosphines are of importance as versatile synthesis intermediates. Thus, dehalogenation of $(tert$-$C_4H_9)_2PCl$ and $tert$-$C_4H_9PCl_2$ with sodium in dioxane leads to tetra-$tert$-butyldiphosphine and tetra-$tert$-butylcyclotetraphosphine (tetra-$tert$-butyltetraphosphetane), respectively.[6] Reduction of $(tert$-$C_4H_9)_2PCl$ with lithium aluminum hydride [lithium tetrahydridoaluminate(1−)] gives the secondary phosphine, $(tert$-$C_4H_9)_2PH$.[7] The reaction of $(tert$-$C_4H_9)_2PCl$ with Grignard reagents gives tertiary phosphines, $RP(tert$-$C_4H_9)_2$. A better route to tertiary phosphines, especially $(tert$-$C_4H_9)_3P$, is available in the reaction of $(tert\text{-}C_4H_9)_2PCl$ with organolithium compounds.[7,14] Tri-*tert*-butylphosphine is of current interest as a ligand.[12]

Both chloro phosphines, $(tert\text{-}C_4H_9)_n PCl_{3-n} (n = 1,2)$, serve as precursors to the fluoro phosphines, $(tert\text{-}C_4H_9)_n PF_{3-n}$, obtained by fluorination of the former by using sodium fluoride in tetrahydrothiophene 1,1-dioxide (sulfolane) or acetonitrile.[11] The addition of sulfur to $(tert\text{-}C_4H_9)_n PCl_{3-n}$ furnished the corresponding sulfides, $(tert\text{-}C_4H_9)_n P(=S)Cl_{3-n}$.[9,11] *tert*-Butyl-phosphonothioic dichloride was also obtained by the reaction of $tert\text{-}C_4H_9P(=O)Cl_2$ with P_4S_{10}.[13,14] Hydrolysis of $tert\text{-}C_4H_9PCl_2$ and $(tert\text{-}C_4H_9)_2PCl$ furnished the acids, $tert\text{-}C_4H_9P(=O)\text{-}$(H)(OH), and $(tert\text{-}C_4H_9)_2P(=O)(H)$.[14] Reaction of $tert\text{-}C_4H_9PCl_2$ with $tert\text{-}C_4H_9Cl\text{--}AlCl_3$ gave the Kinnear-Perren complex which was hydrolyzed to the phosphinic chloride $(tert\text{-}C_4H_9)_2P(=O)Cl$.[14] Amino derivatives have been obtained by the reaction of both $tert\text{-}C_4H_9PCl_2$ and $(tert\text{-}C_4H_9)_2PCl$ with ammonia and amines.[8,15,16] The dichlorophosphine, $tert\text{-}C_4H_9PCl_2$, served as an intermediate in the synthesis of the chloro phosphine, $(tert\text{-}C_4H_9)(CH_3)PCl$.[10]

References

1. Houben-Weyl, "Methoden der Organischen Chemie," Vol. 12, Sec. 1, p. 203, Georg Thieme Verlag, Stuttgart, 1963.
2. R. C. Miller, *J. Org. Chem.*, **24**, 2013 (1959).
3. K. Issleib and W. Seidel, *Chem. Ber.*, **92**, 2681 (1959).
4. W. Voskuil and J. F. Arens, *Rec. Trav. Chim.*, **82**, 302 (1963).
5. S. H. Metzger, O. H. Basedow, and A. F. Isbell, *J. Org. Chem.*, **29**, 627 (1964).
6. K. Issleib and M. Hoffmann, *Chem. Ber.*, **99**, 1320 (1966).
7. H. Hoffmann and P. Schellenbeck, *ibid.*, **100**, 692 (1967).
8. O. J. Scherer and G. Schieder, *ibid.*, **101**, 4148 (1968).
9. G. Hägele, Ph.D. thesis, Technische Hochschule, Aachen, 1969.
10. O. J. Scherer and W. Gick, *Chem. Ber.*, **103**, 71 (1970).
11. M. Fild and R. Schmutzler, *J. Chem. Soc. (A)*, **1970**, 2359.
12. H. Schumann, O. Stelzer, and U. Niederreuther, *J. Organomet. Chem.*, **16**, P64 (1969).
13. P. C. Crofts and I. S. Fox, *J. Chem. Soc. (B)*, **1968**, 1416.
14. P. C. Crofts and D. M. Parker, *ibid. (C)*, **1970**, 332.
15. O. J. Scherer and P. Klusmann, *Angew. Chem.*, **80**, 560 (1968).
16. O. J. Scherer and P. Klusmann, *Z. Anorg. Allgem. Chem.*, **370**, 171 (1969).
17. O. J. Scherer and P. Klusmann, *Angew. Chem.*, **81**, 743 (1969).

3. 1,2-BIS(PHOSPHINO)ETHANE

$$2P(OC_2H_5)_3 + BrC_2H_4Br \xrightarrow{\Delta}$$

$$\underset{\substack{\| \\ O}}{(C_2H_5O)_2P}C_2H_4\underset{\substack{\| \\ O}}{P(OC_2H_5)_2} + 2C_2H_5Br$$

$$2(C_2H_5O)_2\overset{\substack{O \\ \|}}{P}C_2H_4\overset{\substack{O \\ \|}}{P}(OC_2H_5)_2 + 3LiAlH_4 \xrightarrow{Et_2O}$$

$$2H_2PC_2H_4PH_2 + 2LiAl(OC_2H_5)_4 + LiAl(OH)_4$$

Submitted by R. CRAIG TAYLOR* and DOUGLAS B. WALTERS†
Checked by E. R. WONCHOBA‡ and G. W. PARSHALL‡

The importance of ethylenediamine as a chelating agent for metal ions is well established.[1] The compound is an excellent example of a simple bidentate hard base. The synthesis of the phosphorus analog has been reported only recently by Maier.[2] Potentially, 1,2-bis(phosphino)ethane is an excellent example of a simple bidentate soft base; however, very little of its chemistry has been investigated. In particular, no reactions of the ligand with metal ions have been reported. The following detailed procedure is submitted with the hope of stimulating research into the derivative chemistry of this ligand. The procedure outlined here is similar to Maier's, although more details and precautions have been included in order to facilitate the synthesis.

■ *Caution. The lithium aluminum hydride reduction step must be carried out in a good hood because of the extreme toxic nature of 1,2-bis(phosphino)ethane. The product is spontaneously*

*Chemistry Department, Oakland University, Rochester, Mich. 48063.
†Chemistry Department, University of Georgia, Athens, Ga. 30601. This work was supported by the National Science Foundation under grant GP-8512.
‡Central Research Department, E. I. du Pont de Nemours & Company, Wilmington, Del. 19898.

flammable in air; hence, manipulations must be carried out under a nitrogen atmosphere.

Procedure

Although the starting material, tetraethylethylenebis (phosphonate) is commercially available,* its synthesis is reported here. It is possible to prepare this intermediate by a number of different methods;[3-5] however, the Arbusov type of rearrangement[5] is less time-consuming than the others which have been reported.

A mixture of 314 g. (1.89 moles) of triethyl phosphite† and 200 g. (1.05 moles) of 1,2-dibromoethane is placed in a two-necked, 500-ml., round-bottomed flask equipped with a thermometer well. The flask is attached to a Nester-Faust 24-in. intermediate-model spinning-band distillation column with a column bore of 8 mm.‡ The pot temperature is maintained at 145–150°C. After its contents are heated at this temperature for $1\frac{1}{2}$–2 hours, ethyl bromide commences to distill. Approximately 130 ml. (180–190 g., 1.6–1.7 moles) of ethyl bromide is collected. After the removal of ethyl bromide is completed, the reaction mixture is cooled to ambient temperature and subsequently fractionated under reduced pressure. The following fractions are obtained:

$$(C_2H_5O)_2\overset{\overset{O}{\|}}{P}C_2H_5 \qquad\qquad \text{b.p., } 55°\text{C./mm.} \qquad\qquad 100 \text{ g.}$$

$$(C_2H_5O)_2\overset{\overset{O}{\|}}{P}C_2H_4Br \qquad\qquad \text{b.p., } 100\text{–}107°\text{C./mm.} \qquad\qquad 30 \text{ g.}$$

$$(C_2H_5O)_2\overset{\overset{O}{\|}}{P}C_2H_4\overset{\overset{O}{\|}}{P}(OC_2H_5)_2 \qquad \text{b.p., } 155\text{–}157°\text{C./mm.} \qquad 125 \text{ g., } 44\% \text{ yield}\S$$

---

*Strem Chemicals, Inc., 150 Andover St., Danvers, Mass. 01923.

†Although the checkers suggested freshly distilled triethyl phosphite, the authors have found the material obtained from Aldrich is quite satisfactory.

‡Any efficient fractionating column may be used in place of the described one. However, the checkers found that the bore of the fractionating column was quite critical for the success of the synthesis.

§Checkers obtained 71.6 g. (25%).

The product obtained in this fashion is suitable for the next step in the synthesis.

The following reaction must be carried out under a nitrogen atmosphere. Extreme caution should be exercised during the collection of the product because of its toxic and flammable nature.

A 3-l., three-necked, round-bottomed flask is equipped with an air-driven mechanical stirrer, a 250-ml. pressure-equalizing dropping funnel with a nitrogen inlet attached at the top, and a Freiderich's condenser. The gases emanating from the condenser outlet are allowed to pass through an empty 1-l. flask and subsequently through a series of two 250-ml. Drechsel gas-washing bottles fitted with fritted disks and half-filled with bromine water.* This precaution effectively traps the toxic vapors produced during the reduction step.

The entire system is flushed thoroughly with dry nitrogen before charging with reactants. A suspension of 40 g. (1.05 moles) of lithium aluminum hydride in 1 l. of anhydrous diethyl ether is introduced into the flask. The dropping funnel is charged with a solution of 100 g. (0.33 mole) of tetraethylethylenebis(phosphonate) in 180 ml. of diethyl ether. The flask and its contents are cooled with an external petroleum ether (100–115°C.)–Dry Ice bath maintained at 0°C. The solution containing the phosphonate is added dropwise with stirring during the course of 3 hours.

(■ *Caution. A cracked flask could be dangerous.*) The reaction mixture is allowed to stand overnight at ambient temperature, after which it is hydrolyzed by the slow addition of 800 ml. of 6 N hydrochloric acid. The ether layer is separated and dried over anhydrous sodium sulfate for 8 hours. The dried ether layer is transferred, under nitrogen, to a 1-l., round-bottomed flask with 24/25 and 10/30 standard-taper (S.T.) outer joints. The mixture is fractionated under a slight positive pressure of nitrogen with the use of the still and traps previously described. The ether is

*The checkers found that full-strength Clorox is a satisfactory substitute for bromine water.

removed by gentle heating. The temperature is raised gradually, and ethyl alcohol (a product of the reduction) distills at 78°C. Finally, the 1,2-bis(phosphino)ethane distills at 114–117°C. (in literature,[2] 114–117°C./725 torr) as a colorless liquid. The yield is 18 g. (57%).* It is convenient for future handling purposes to collect the product in previously weighed glass ampuls which can be sealed off rapidly with a gas-oxygen torch. The use of stopcocks should be avoided, for the product attacks most commonly used laboratory greases.

Properties

1,2-Bis(phosphino)ethane is a colorless, volatile liquid with the very unpleasant odor characteristic of primary and secondary phosphines. Its vapor pressure[2] may be represented by the equation:

$$\log P \text{ (mm.)} = -\frac{1988}{T} + 8.02813 \qquad T = \text{kelvin}$$

It is immiscible in water but is very soluble in common organic solvents, e.g., ethanol, ether, tetrahydrofuran, and benzene. The gaseous infrared spectrum[2] exhibits a strong P–H stretch at 2292 cm.$^{-1}$. Other infrared bands occur at 2920, 1085, 930, 838, 820, and 670 cm.$^{-1}$. The ^{31}P n.m.r. spectrum shows the expected triplet centered at 131 p.p.m., with a J_{P-H} of 193 Hz.† The ^{1}H n.m.r. spectrum consists of a doublet centered at 3.3 p.p.m., with J_{P-H} = 194 Hz. Each half of this doublet is further split into a triplet as a result of the coupling of the phosphorus protons with the methylene protons ($J \approx 6$ Hz.). The methylene protons exhibit a complex pattern centered at 2.22 p.p.m.‡

*The checkers obtained a yield of 16 g.

†The ^{31}P n.m.r. spectrum was run at 24.288 MHz. on a Hitachi–Perkin Elmer R-20 with 85% H_3PO_4 as an external standard.

‡The ^{1}H n.m.r. was run at 60 MHz. on the same instrument with TMS as an internal standard.

References

1. A. E. Martell and M. Calvin, "Chemistry of Metal Chelate Compounds," Prentice-Hall, Inc., New York, 1952.
2. L. Maier, *Helv. Chim. Acta,* 49, 842 (1966).
3. A. M. Kinnear and E. A. Perren, *J. Chem. Soc.,* 1952, 3437.
4. K. Moedritzer and R. R. Irani, *J. Inorg. Nucl. Chem.,* 22, 297 (1961).
5. A. H. Ford-Moore and J. H. Williams, *J. Chem. Soc.,* 1947, 1465.

4. TETRAMETHYLDIPHOSPHINE AND FLEXIBLE ALIPHATIC (DIMETHYLPHOSPHINO) LIGANDS

Submitted by G. KORDOSKY,* B. R. COOK,* JOHN CLOYD, JR.,*
and DEVON W. MEEK*
Checked by G. W. PARSHALL† and E. R. WONCHOBA†

Simple trialkylphosphines, R_3P, can be prepared relatively easily by using excess Grignard or alkyllithium reagents.[1,2] However, the synthesis of mixed $R_2R'P$ monophosphines or chelating aliphatic ligands of the type of $R_2P(CH_2)_nA$ (where A = a donor group and $n = 2,3,4$) have been almost prohibitively difficult, owing to the toxicity, chemical reactivity, oxidation sensitivity, and commercial unavailability of likely precursors.

Recent use of phosphine ligands in the areas of catalysis,[3,4] pentacoordination,[5,6] and oxidative-addition reactions[7] has prompted investigations into variations of donor basicity, chelate chain length, and mixed sets of donor atoms. Thus, relatively accessible synthetic routes to chelating aliphatic phosphines are needed. Removal of sulfur from a diphosphine disulfide[8] and subsequent preparation of $NaPR_2$ in liquid ammonia provides a useful laboratory route to aliphatic phosphines. For example,

*Ohio State University, Columbus, Ohio 43210.
†Central Research Department, E. I. du Pont de Nemours & Company, Wilmington, Del. 19898.

tetramethyldiphosphine disulfide, $(CH_3)_2\underset{\underset{S}{\|}}{P}-\underset{\underset{S}{\|}}{P}(CH_3)_2$, and tetra-

ethyldiphosphine disulfide, $(C_2H_5)_2\underset{\underset{S}{\|}}{P}-\underset{\underset{S}{\|}}{P}(C_2H_5)_2$, are commer-

cially available; in addition, the tetrapropyl-, tetrabutyl-, and bis(methylphenyl)diphosphine disulfides are available by a Grignard reaction.[9]

Preparations of the following three aliphatic chelating (dimethylphosphino) ligands illustrate the synthetic route and the procedure that can lead to a large number of new phosphine ligands.

A. TETRAMETHYLDIPHOSPHINE

$$(CH_3)_2\underset{\underset{S}{\|}}{P}-\underset{\underset{S}{\|}}{P}(CH_3)_2 + 2(n\text{-}C_4H_9)_3P \longrightarrow$$

$$(CH_3)_2P-P(CH_3)_2 + 2(n\text{-}C_4H_9)_3P{=}S$$

Procedure

■ *Caution. Tetramethyldiphosphine and dimethylphosphine are toxic and spontaneously flammable in air. These reactions should be carried out in a purified nitrogen atmosphere and in a very good hood.*

A carefully dried, 500-ml., three-necked, round-bottomed flask is fitted with a compact distillation head and condenser, a pressure-equalizing addition funnel, and a nitrogen gas inlet tube. The nitrogen passes out through the distillation head. If conventional stopcocks are used, they should be lubricated with fluorocarbon or silicone grease.

Tetramethyldiphosphine disulfide (46.5 g., 0.25 mole) and 111 g. (0.50 mole plus 10% excess) of tributylphosphine are added directly to the three-necked flask.* The funnel is replaced with a

*Tetraethyl- and tetramethyldiphosphine disulfide (Orgmet, Inc., 300 Neck Rd. Haverhill, Mass. 01830), tributylphosphine (Carlisle Chemical Works, Inc., Reading, Ohio 45215), and 1,3-dichloropropane (Eastman Kodak Company, Rochester, N.Y. 14600) were obtained commercially and used without further purification. The first two reactants are completely inert to one another at room temperature and can be placed in the reaction vessel simultaneously.

glass stopper, and the temperature is raised slowly with an oil bath to 170°C. At this point, the tetramethyldiphosphine disulfide dissolves and reaction commences, as indicated by a gentle frothing of the solution. The reaction is essentially complete within $\frac{1}{2}$ hour, after which the frothing subsides. The reaction mixture is cooled to room temperature, and the distillation head is connected to a vacuum distillation setup. The three-necked flask is heated in an oil bath to distill tetramethyldiphosphine at 61–63°C. at 55 torr.* The yield is 24.4–26.8 g. (80–88%).†

Since tetramethyldiphosphine is so sensitive to air, the following procedure is useful for detaching the collection flask from the distillation head and for obtaining the weight of the sample. The nitrogen flow is continued while the flask that contains the tetramethyldiphosphine is cooled to −78°C. (At −78°C. the diphosphine is a solid.) Concurrently, a glass, hollow stopper is fitted with a septum cap that contains a small syringe needle (*A* in Fig. 2). The needle is attached by Tygon tubing, placed between two gas bubblers, and connected to the dry nitrogen gas. This arrangement ensures that a stream of dry nitrogen passes through the needle when it is not attached to a flask; also it ensures a positive nitrogen pressure inside the flask. The flask containing the solid tetramethyldiphosphine is removed from the distillation apparatus, and the syringe-needle glass stopper is quickly attached. After the flask warms to room temperature, the needle is removed from the septum, the flask is weighed, and the weight of the tetramethyldiphosphine is determined by difference. The needle and nitrogen stream are reinserted into the septum, and the flask is cooled again to −78°C. The syringe-needle glass stopper is removed, and the flask is attached by a curved glass adapter (*B* in Fig. 2) to an addition funnel, preparatory to formation of sodium dimethylphosphide [(dimethylphosphino)sodium]. The material

---

*Tetramethyldiphosphine distills at 131°C. at *ca.* 750 torr. However, the vacuum distillation is preferred for safety; also less frothing occurs at the lower temperature.

†Tributylphosphine sulfide, $(n\text{-}C_4H_9)_3$ P=S, remains in the three-necked flask, since it distills at 111°C./0.1 torr.

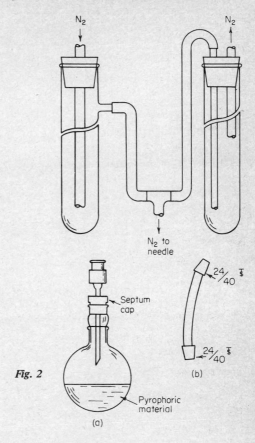

Fig. 2

that has been collected as above is used in the three syntheses below without additional purification.

B.   1,3-BIS(DIMETHYLPHOSPHINO)PROPANE

$$[(CH_3)_2 P]_2 + 2Na \longrightarrow 2NaP(CH_3)_2$$
$$2(CH_3)_2 PNa + ClCH_2 CH_2 CH_2 Cl \longrightarrow$$
$$(CH_3)_2 PCH_2 CH_2 CH_2 P(CH_3)_2 + 2NaCl$$

Procedure

A 500-ml., three-necked, round-bottomed flask is charged with 300 ml. of anhydrous liquid ammonia and equipped with a Dry

Ice–acetone condenser, a mechanical stirrer, a nitrogen gas inlet-outlet system, and the pressure-equalizing addition funnel containing the tetramethyldiphosphine. The liquid ammonia is maintained at $-78°$C. during the reactions. Freshly cut pieces of sodium (8.4 g., 0.37 mole) are added to the liquid ammonia, which produces a blue solution. The tetramethyldiphosphine (22.6 g., 0.185 mole) is added dropwise. A deep-green solution develops as the diphosphine is added. The addition funnel is rinsed with 25 ml. of anhydrous diethyl ether to remove the remaining tetra-methyldiphosphine, and it is then recharged with 20.7 g. (0.18 mole) of 1,3-dichloropropane in 50 ml. of ether. This solution is added dropwise, whereupon the color of the liquid-ammonia solution becomes bright red. Addition of the 1,3-dichloropropane solution is stopped when the liquid ammonia–$(CH_3)_2$ PNa solution becomes colorless, since the excess would react with the product to produce phosphonium salts. The cold bath and CO_2 condenser are removed and the ammonia is allowed to evaporate into a hood. When the flask reaches room temperature, 200 ml. of dry ether is added. The resulting slurry, which contains the liquid bis(phosphine) and the precipitated sodium chloride, is transferred to a closed filtering flask containing 5–6 g. of anhydrous sodium sulfate (A in Fig. 3). The filtering flask is attached to a 50-ml., three-necked, round-bottomed flask which is equipped with a micro distillation head, short Vigreux column, and nitrogen inlet tube. The ether is removed and the 1,3-bis(dimethylphosphino)propane is distilled at $62-63°$C./4 torr. The yield is *ca.* 24.3 g. (80%).

Properties

1,3-Bis(dimethylphosphino)propane is a colorless liquid that is very sensitive to air and other oxidizing agents. A proton n.m.r. spectrum of the neat liquid has two doublets centered at 9.06τ and 8.57τ (J_{P-H} = 2.9 Hz.), which are assigned to the methyl and methylene protons, respectively, on the basis of the 2.0:1.0 integration.

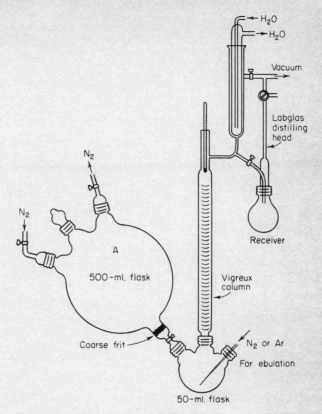

Fig. 3. Distillation apparatus for the ligands.

The disulfide derivative of the ligand is prepared by adding 0.23 g. of recrystallized sulfur to 0.6 g. of the diphosphine in 50 ml. of dry benzene. An exothermic reaction occurs, and a white solid is deposited. The disulfide is recrystallized from dichloromethane; m.p. = 226–227.5°C. The infrared P=S stretching frequency occurs at 575 cm.$^{-1}$. *Anal.* Calcd. for $C_7H_{18}P_2S_2$: C, 36.83; H, 7.95; S, 28.09. Found: C, 36.25; H, 8.01; S, 28.74.

The bis(phosphine) forms stable complexes with a large number of transition metals. For example, the five-coordinate $[ML_2X]^+$ (M = Ni, Co) and the six-coordinate *trans*-$[CoL_2X_2]^+$ complexes have been characterized.[10]

C.   (1-DIMETHYLPHOSPHINO)(3-DIMETHYLARSINO)-PROPANE

$$(CH_3)_2 PNa + ClCH_2 CH_2 CH_2 As(CH_3)_2 \longrightarrow$$
$$(CH_3)_2 PCH_2 CH_2 CH_2 As(CH_3)_2 + NaCl$$

Procedure

A sodium dimethylphosphide solution is prepared as in Sec. B from 4.6 g. (0.20 mole) of sodium, 200 ml. of anhydrous liquid ammonia, and 12.2 g. (0.10 mole) of tetramethyldiphosphine. Then 36.5 g. (0.20 mole) of (3-chloropropyl)dimethylarsine[11] is added dropwise to the reaction mixture. The color of the reaction mixture changes slowly from green to red and finally becomes colorless when the addition is complete. The liquid ammonia is allowed to evaporate into a hood at room temperature, and the remaining oil is distilled under vacuum; b.p. 43–44°C./torr. The yield is *ca.* 26 g. (0.125 mole, 63%).

Properties

(1-Dimethylphosphino)(3-dimethylarsino)propane is a colorless liquid that is sensitive to air and other oxidizing agents. A proton n.m.r. spectrum of the neat liquid exhibits a singlet at 9.12τ, a doublet at 9.07τ (J_{P-H} = 3 Hz.), and a complex multiplet centered at 8.52τ, which are assigned the As–CH_3, P–CH_3, and methylene protons, respectively.

The disulfide of $(CH_3)_2 PCH_2 CH_2 CH_2 As(CH_3)_2$ is prepared by adding 0.005 mole of the ligand to 0.32 g. (0.01 mole) recrystallized sulfur in 25 ml. of warm benzene. The crystals, which begin to appear almost immediately, are recrystallized from ethanol and dried *in vacuo*. The yield is 1.24 g. (91%); m.p., 234–236°C. *Anal.* Calcd. for $C_7 H_{18} AsPS_2$: C, 30.88; H, 6.62; S, 23.53. Found: C, 30.60; H, 6.68; S, 23.79. The mass spectrum of the disulfide shows the parent ion peak at m/e = 272 and major peaks at 225 ($C_6 H_{15} PAsS^+$), 135 (($CH_3)_2 PCH_2 CH_2 CH_2^+$) and 93

$((CH_3)_2 \overset{\|}{\underset{S}{P}}{}^+)$. The infrared P=S and As=S stretching frequencies are at 555 and 466 cm.$^{-1}$, respectively.

The bis(methyl iodide) quaternary salt of the arsenic phosphorus ligand is prepared by adding the ligand (0.005 mole in 5 ml. of ethanol) to a 50-ml., round-bottomed flask containing 20 ml. of methyl iodide. Immediately a white precipitate forms, which is stirred for 20 minutes at room temperature and then collected on a filter. The white solid is recrystallized from hot ethanol by adding ether until a slight turbidity is apparent. The mixture is then cooled slowly, and the resulting crystals are collected and dried *in vacuo*. Yield, 2.4 g. (97%); m.p. 259–260°C. *Anal.* Calcd. for $C_9 H_{24} AsI_2 P$: C, 21.95; H, 4.88; I, 51.62. Found: C, 21.91; H, 4.84; I, 51.37. The above salt exhibits a divalent electrolyte conductance behavior in nitromethane.

(1-Dimethylphosphino)(3-dimethylarsino)propane forms stable complexes with a large number of transition metals; the four-coordinate $NiLX_2$, the five-coordinate $[CoL_2 X]^+$, and the six-coordinate $[RhL_2 X_2]^+$ complexes are illustrative of the types that have been characterized.[12,13]

D. (1-DIMETHYLPHOSPHINO)(3-DIMETHYLAMINO)-PROPANE

$$2ClCH_2 CH_2 CH_2 N(CH_3)_2 \cdot HCl + Na_2 CO_3 \longrightarrow$$
$$2ClCH_2 CH_2 CH_2 N(CH_3)_2 + H_2 O + CO_2 + 2NaCl$$
$$(CH_3)_2 PNa + ClCH_2 CH_2 CH_2 N(CH_3)_2 \longrightarrow$$
$$(CH_3)_2 PCH_2 CH_2 CH_2 N(CH_3)_2 + NaCl$$

Procedure

■ *Caution. The free amine is related to nitrogen mustards and must be handled carefully in an efficient hood.*

Freshly distilled 3-chloro-*N,N*-dimethylpropylamine* (21.5 g., 0.18 mole) is added dropwise to a sodium dimethylphosphide

---

*Commercial 3-chloro-*N,N*-dimethylpropylamine hydrochloride (Michigan Chemical Corporation, 351 E. Ohio St., Chicago, Ill. 60611) was added to cold (0°C.), saturated solution that is prepared as in Sec. B from 10.8 g. (0.09 mole) of tetramethyldiphosphine. During the addition of 3-chloro-N,N-dimethylpropylamine, the color of the reaction mixture changes from green to red to orange to a white turbid suspension. The liquid ammonia is allowed to evaporate into the hood, and the resulting oil is extracted from the sodium chloride with the three 75-ml. portions of anhydrous deaerated ether.* The ether extract is dried over anhydrous sodium sulfate for 2 hours, and then the ether is removed by distillation. The oil that remains is vacuum-distilled at 36°C./3 torr.† The yield is 16.7 g. (68%).

Properties

(1-Dimethylphosphino)(3-dimethylamino)propane is a colorless liquid that is sensitive to air oxidation. The proton n.m.r. spectrum of the neat compound has a six-proton doublet (J_{P-H} = 2.5 Hz.) at 9.07τ, a four-proton multiplet centered at 8.60τ, a six-proton singlet at 7.88τ, and a two-proton multiplet at 7.77τ, which are assigned to the $\diagup$P–CH$_3$, $\diagup$PCH$_2$CH$_2$-, $\diagup$N–CH$_3$, and N–CH$_2$- protons, respectively. The mass spectrum of the ligand shows the parent ion at m/e = 147 and major peaks at m/e = 132 ($C_6H_{15}NP^+$), 89 ((CH_3)$_2$PCH$_2$CH$_2^+$), 75 ((CH_3)$_2$PCH$_2^+$), 61 ((CH_3)$_2$P$^+$), 58 ((CH_3)$_2$NCH$_2^+$), and 44 ((CH_3)$_2$N$^+$).

The bis(methyl iodide) derivative of (CH_3)$_2$PCH$_2$CH$_2$CH$_2$N-(CH_3)$_2$ is prepared by adding 0.5 g. of the compound to a 50-ml. flask containing 20 ml. of stirred methyl iodide. The white solid, which appears immediately, is stirred for 2 hours, collected on a filter funnel, and recrystallized two times from absolute ethanol, m.p., 270–271°C. *Anal.* Calcd. for $C_9H_{24}I_2NP$: C, 25.08; H, 5.61; I, 58.87. Found: C, 25.35; H, 5.49; I, 58.67.

The white salt is a divalent electrolyte in nitromethane, and

---

sodium carbonate solution covered with ether. After stirring for 5 minutes, the ether layer containing the free amine was separated, dried over potassium carbonate, and 3-chloro-N,N-dimethylpropylamine distilled at 65°C. at 68 torr.

*The sodium chloride may also be removed by filtration with the filtering flask, as shown in Fig. 3.

†The checkers report a b.p. of 75–76°C. at 26 torr.

its proton n.m.r. spectrum in D_2O has a doublet at 8.02τ (J_{P-H} = 14 Hz.), a singlet at 6.78τ, a complicated multiplet centered at 7.80τ, and a complicated multiplet centered at 6.48τ. The doublet and singlet are assigned easily to the $(CH_3)_3P^+$- and $(CH_3)_3N^+$- groups, respectively.

References

1. J. R. Van Wazer, "Phosphorus and Its Compounds," Interscience Publishers, Inc., New York, 1958.
2. L. Maier, in "Progress in Inorganic Chemsitry," F. A. Cotton (ed.), Vol. 5, pp. 17–210, Interscience Publishers, Inc., New York, 1963.
3. J. A. Osborn, F. H. Jardine, J. F. Young, and G. Wilkinson, *J. Chem. Soc. (A)*, **1966**, 1711.
4. B. R. James and F. T. T. Ng, *Chem. Commun.*, **1970**, 908.
5. E. C. Alyea and D. W. Meek, *J. Am. Chem. Soc.*, **91**, 5761 (1969).
6. L. M. Venanzi, *Angew. Chem., Intern. Ed. Engl.*, **3**, 543 (1964).
7. M. A. Bennett and D. L. Milner, *J. Am. Chem. Soc.*, **91**, 6983 (1969).
8. G. W. Parshall, *J. Inorg. Nucl. Chem.*, **14**, 291 (1960); L. Maier, *ibid.*, **24**, 275 (1962).
9. N. K. Patel and H. J. Harwood, *J. Org. Chem.*, **32**, 2999 (1967).
10. J. Cloyd, Jr., Ph.D. dissertation, The Ohio State University, Columbus, Ohio, 1970.
11. G. S. Benner, M.S. thesis, The Ohio State University, 1963; G. S. Benner, W. E. Hatfield, and D. W. Meek, *Inorg. Chem.*, **3**, 1544 (1964).
12. B. R. Cook, Ph.D. dissertation, The Ohio State University, Columbus, Ohio, 1972.
13. G. Kordosky, Ph.D. dissertation, The Ohio State University, Columbus, Ohio, 1971.

5. AMIDO(PHOSPHONITRILIC CHLORIDE—CYCLIC TRIMER) AND 1,1-DIAMIDO(PHOSPHONITRILIC CHLORIDE— CYCLIC TRIMER)

Submitted by GERALD R. FEISTEL,* MARLENE K. FELDT,† RONALD L. DIECK,‡ and THERALD MOELLER‡
Checked by GARY L. LUNDQUIST § and C. DAVID SCHMULBACH §

Only three amido derivatives of phosphonitrilic chloride—cyclic trimer have been described, namely, the compounds

*Nalco Chemical Company, 6225 W. 66th Pl., Chicago, Ill. 60638.
†Highland Drive, Route 1, Belle Mead, N.J. 08502.
‡Arizona State University, Tempe, Ariz. 85281.
§Southern Illinois University, Carbondale, Ill. 62901.

$N_3P_3Cl_5(NH_2)$, $1,1\text{-}N_3P_3Cl_4(NH_2)_2$, and $N_3P_3(NH_2)_6$. The di-amido compound has been prepared by the reaction of an ethereal solution of phosphonitrilic chloride—cyclic trimer with either gaseous[1-3] or aqueous[1-7] ammonia. No other ammonolysis product has been isolated from either reaction. The aqueous-ammonia procedure has the advantages of ease of handling and larger yield. The monoamido compound was first obtained, but only in small yield, as a pyrolysis product of the diamido derivative.[3] However, the controlled removal of a single amido group from the diamido compound by using gaseous hydrogen chloride gives a larger yield of pure product.[4,5] The hexaamido derivative is a product of the exhaustive ammonolysis of the trimeric cyclic chloride with liquid ammonia.[8,9]

A. 1,1-DIAMIDO(PHOSPHONITRILIC CHLORIDE—CYCLIC TRIMER)

$$N_3P_3Cl_6 + 4NH_3 \longrightarrow 1,1\text{-}N_3P_3Cl_4(NH_2)_2 + 2NH_4Cl$$

Procedure

One hundred fifty grams (0.43 mole) of phosphonitrilic chloride—cyclic trimer[10] is placed in a single-necked, 2-l., round-bottomed flask which is equipped with a reflux condenser and a magnetic stirrer. Diethyl ether (500 ml.) is added to dissolve the solid. Four hundred and fifty grams of 15 M aqueous ammonia is added slowly through the reflux condenser while the contents of the flask are being stirred. After addition is complete, the mixture is heated at reflux for one hour while stirring is continued. The resulting two-phase system is transferred to a separatory funnel, where the ether layer is removed and washed several times with water. The ether solution is then dried over anhydrous calcium chloride, which is followed by solvent removal with a rotary evaporator. The resulting white solid is heated with 200 ml. of n-heptane to dissolve unreacted phosphonitrilic chloride—cyclic trimer. (The checkers report that the product at this stage melts at

163–165°C.) The solid remaining after filtration of the cooled suspension is recrystallized from diethyl ether by cooling the saturated solution to −70°C. Yield, based upon phosphonitrilic chloride—cyclic trimer, is 80–95 g. (60–72%); m.p., 163–165°C. *Anal.* Calcd. for $N_3P_3Cl_4(NH_2)_2$: N, 22.68; H, 1.31. Found: N, 22.94; H, 1.35. The checkers report equally good results, working on one-third of the above scale.

Properties

1,1-Diamido(phosphonitrilic chloride—cyclic trimer) is a white, crystalline substance that melts over the range of 163–165°C. The compound is soluble in diethyl ether. Thermal decomposition is appreciable at room temperature but slow at −10°C. That this procedure yields only a single compound suggests that the substance has the 1,1, or geminal, molecular structure, rather than the 1,3 structure that allows for cis-trans isomerism.[4,5] Although [31]P n.m.r. data for the compound itself do not distinguish between the 1,1 and 1,3 configurations, comparable data for the compound $N_3P_3Cl_4(N=PCl_3)_2$, formed as a product of the reaction of the diamido compound with phosphorus(V) chloride,[4-7] confirm the 1,1 arrangement of amido groups.[4,5,11] Nuclear magnetic resonance data for phenylated derivatives of $N_3P_3Cl_4(N=PCl_3)_2$[6,7] and data for phenoxy derivatives of the diamido compound[12] also support the geminal molecular structure. Group chemical shifts in the [31]P n.m.r. spectrum are PCl_2, −18.3 p.p.m., and $P(NH_2)_2$, −9.03 p.p.m. (both versus 85% H_3PO_4).[4,5] Important bands in the infrared spectrum are 3340 and 3240(vs) (N—H stretching), 1535(s) (N—H deformation), *ca.* 1200(vs) (P—N ring stretching), 825(s) (P—Cl_2 vibration), 650(m), 600(vs), and 518(vs) cm.$^{-1}$.[6]

B. AMIDO(PHOSPHONITRILIC CHLORIDE— CYCLIC TRIMER)

$$N_3P_3Cl_4(NH_2)_2 + 2HCl \longrightarrow N_3P_3Cl_5(NH_2) + NH_4Cl$$

Procedure

One hundred grams (0.32 mole) of 1,1-diamido(phosphonitrilic chloride—cyclic trimer) is placed in a 500-ml., three-necked flask and dissolved therein in 250 ml. of freshly distilled 1,4-dioxane.* A reflux condenser and magnetic stirrer are attached. The flask is equipped with a gas inlet tube, and the apparatus and its contents are weighed. Hydrogen chloride gas is bubbled through the solution. The flask, its contents, and the inlet tube are weighed frequently, and the flow of hydrogen chloride is terminated when a stoichiometric quantity has been added (total increase in weight = 23.4 g.).† The solution is heated under reflux and with constant stirring for 8 hours. (Checkers found $6\frac{1}{2}$ hours to be adequate.) The hot suspension is filtered to remove ammonium chloride. The yellow filtrate is freed of solvent with a rotary evaporator. The resulting brown solid-oil mixture (checkers report off-white solid) is treated with 250 ml. of boiling carbon tetrachloride. The resulting solution decolorizes (imperfectly) under reflux with activated charcoal. The charcoal is removed by filtration through diatomaceous earth. The carbon tetrachloride is removed by careful evaporation to yield the crystalline product. Yield, based upon the diamido compound, is 70–76 g. (67–73%); m.p., 138–139°C. *Anal.* Calcd. for $N_3P_3Cl_5(NH_2)$: N, 17.06; P, 28.30; Cl, 54.00; H, 0.61. Found: N, 17.01; P, 28.41; Cl, 53.08; H, 0.65. (Checkers report excellent results on a $\frac{1}{5}$ scale in quantity.)

Properties

Amido(phosphonitrilic chloride—cyclic trimer) is a white crystalline, slightly hygroscopic solid, which melts at 138–139°C. and dissolves in both 1,4-dioxane and carbon tetrachloride. Slow

*The checkers found that removing peroxides by passing dioxane through an activated alumina column prior to distillation greatly reduced discoloration of the reaction solution which develops during reflux.

†An excess of hydrogen chloride decreases the yield markedly.

decomposition on standing at room temperature releases hydrogen chloride. It reacts readily with phosphorus(V) chloride to yield the (trichlorophosphazo) derivative $N_3P_3Cl_5(N=PCl_3)$.[4-6] Group chemical shifts in the ^{31}P n.m.r. spectrum are PCl_2, -20.4 p.p.m., and $PCl(NH_2)$, -19.0 p.p.m. (both versus 85% H_3PO_4).[4,5] Important bands in the infrared spectrum are 3200(vs) (N–H stretching), 1535(vs) (N–H deformation), 1200(vs) (P–N ring stretching), 979(s) ($P-NH_2$ stretching), 855(s) ($P-Cl_2$ vibration), and 740(s) cm.$^{-1}$.[6] (Checkers report 3350, 3250, 1542, 1200, 988, 860 and 848 doublet, 750, 600, 528 cm.$^{-1}$.)

References

1. H. Moureu and A. M. de Fiquelmont, *Compt. Rend.*, **198**, 1417 (1934).
2. H. Moureu and P. Rocquet, *Bull. Soc. Chim. France*, **3**, 821 (1936).
3. A. M. de Fiquelmont, *Ann. Chim.*, **12**, 169 (1939).
4. G. R. Feistel, doctoral dissertation, University of Illinois, Urbana, Ill., 1965.
5. G. R. Feistel and T. Moeller, *J. Inorg. Nucl. Chem.*, **29**, 2731 (1967).
6. M. K. Feldt, doctoral dissertation, University of Illinois, Urbana, Ill., 1967.
7. M. K. Feldt and T. Moeller, *J. Inorg. Nucl. Chem.*, **30**, 2351 (1968).
8. L. F. Audrieth and D. B. Sowerby, *Chem. & Ind. (London)*, 748 (1959).
9. D. B. Sowerby and L. F. Audrieth, *Chem. Ber.*, **94**, 2670 (1961).
10. M. L. Nielsen and G. Cranford, *Inorganic Syntheses*, **6**, 94 (1960).
11. W. Lehr, *Z. Anorg. Allgem. Chem.*, **350**, 18 (1967).
12. E. T. McBee, K. Okuhara, and C. J. Morton, *Inorg. Chem.*, **5**, 450 (1966).

Chapter Two

NON-TRANSITION-METAL COMPOUNDS

6. PERCHLORYL FLUORIDE

$$KClO_4 + 2HF + SbF_5 \longrightarrow ClO_3F + KSbF_6 + H_2O$$

Submitted by C. A. WAMSER,* B. SUKORNICK,† W. B. FOX,‡
and D. GOULD§
Checked by A. F. CLIFFORD¶ and J. W. THOMPSON¶

Perchloryl fluoride, ClO_3F, has been prepared by the action of elemental fluorine on potassium perchlorate,[1] by the electrolysis of sodium perchlorate in anhydrous hydrogen fluoride,[2] and by reactions of various metal perchlorates with fluorosulfuric acid[3] or antimony(V) fluoride.[4] Although the latest methods (with HSO_3F or SbF_5 or both) are more convenient than the first ones, they suffer the disadvantages that elevated temperatures are

*Allied Chemical Corporation, Industrial Chemicals Division, Syracuse, N.Y. 13209.
†Allied Chemical Corporation, Specialty Chemicals Division, Buffalo, N.Y. 14200.
‡U.S. Naval Research Laboratory, Washington, D.C. 20390.
§Allied Chemical Corporation, Corporate Chemical Research Laboratories, Morristown, N.J. 07960.
¶Virginia Polytechnic Institute and State University, Blacksburg, Va. 24061.

required, yields are poor, and secondary reactions give impure products. A recent study of the solvolysis of perchlorates in fluorinated solvents[5] has led to the development of a method more conveniently suited to the laboratory preparation of perchloryl fluoride. The method is based on the reaction of potassium perchlorate with the highly acidic system $HF-SbF_5$, which produces perchloryl fluoride in good yield and reasonable purity under mild conditions.

The new compound perbromyl fluoride, BrO_3F, was recently successfully synthesized in high yield by an adaptation of this procedure with potassium perbromate used in place of the perchlorate.[6]

Procedure

▪ *Caution. Perchloryl fluoride is moderately toxic and has strong oxidant properties. Anhydrous hydrogen fluoride and antimony pentafluoride are hazardous, and proper precautions should be observed in their handling. The solution of antimony pentafluoride in anhydrous hydrogen fluoride is one of the strongest acids known. It is an extremely corrosive mixture, not only because of its acidity but also because of its powerful fluorinating action. In addition, the solution containing potassium perchlorate is a very powerfully oxidizing mixture, and every care should be taken to prevent contact with organic materials. Precautions should be taken to avoid not only the liquid hydrogen fluoride itself, but also jets of hydrogen fluoride vapor which may escape through leaks in the equipment. Burned areas of skin should be immersed in lukewarm running water immediately upon detection and be kept completely immersed until any whitish areas which have developed go away. The burns should then be covered with a dry gauze bandage. Do not treat with lime water or other bases, which merely aggravate the burn. Calcium gluconate injections should be used only as a last resort.*

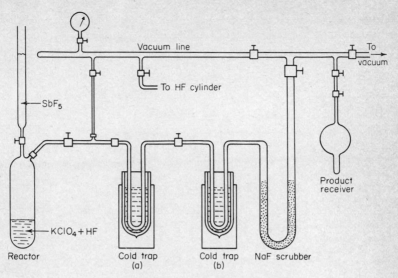

Fig. 4. Apparatus for the preparation of perchloryl fluoride.

The apparatus consists of a simple vacuum manifold similar to that shown in Fig. 4. The vacuum manifold should be fabricated from nickel or Monel metal, with silver-soldered or Swagelok connections. Valves should be of Monel metal, and the traps and scrubber, the function of which is to remove hydrogen fluoride vapor from the ClO_3F product, may be constructed conveniently of poly(chlorotrifluoroethylene) (Kel-F) tubing. A nickel cylinder of 500-ml. or 1-l. capacity serves as a reaction vessel, and a 500-ml. Monel bulb or similar vessel may be used as a product receiver. The scrubber* (about 1-in. o.d.) is charged with pellets of sodium fluoride prepared in a separate operation by heating sodium

---

*The checkers replaced the scrubber section with a 500-ml., 316 stainless-steel, Whitney double-ended cylinder, which allowed the ClO_3F to pass through more slowly and allowed it to be collected free of hydrogen fluoride. The cylinder was wrapped with heating tape, which need not be used during reactions unless large amounts of hydrogen fluoride are to pass through the cylinder. In this case the temperature should be kept at about 95°C. The tape was used to heat the cylinder to 250–300°C. with nitrogen passing through to regenerate the sodium fluoride for subsequent runs.

hydrogen fluoride pellets* in an iron tube to 250–300°C. in a stream of dry air or nitrogen until hydrogen fluoride ceases to be evolved.

For the preparation of about 1-l. of gaseous perchloryl fluoride (1 atmosphere, ambient temperature), 6 g. (0.045 mole) of potassium perchlorate is introduced into the reactor. The reactor is connected to the apparatus as shown and provided with a poly(chlorotrifluoroethylene) (Kel-F) reservoir containing about 4–5 ml. of liquid antimony(V) fluoride. The reactor is evacuated and cooled to about −40°C. Then about 50 g. of anhydrous hydrogen fluoride is condensed onto the potassium perchlorate in the vessel by conventional vacuum-line techniques. After closing the reactor valve, the $KClO_4$–HF mixture is allowed to warm to room temperature with occasional agitation to promote dissolution of most of the potassium perchlorate (solubility of potassium perchlorate in hydrogen fluoride is 9.7% by weight at 0°C.). The reactor is cooled to −40°C., and about 4 ml. of the liquid antimony(V) fluoride is drawn into the vessel. With the reactor inlet and outlet valves both closed, the reaction mixture is warmed to room temperature for one hour. Yields of 60–70% perchloryl fluoride (based on potassium perchlorate) are normally obtained; increases in yield to 85–90% perchloryl fluoride may be realized if the reaction mixture is warmed to 40–50°C. After reaction, the mixture is cooled to −30°C. to reduce the vapor pressure of hydrogen fluoride prior to transferring the product perchloryl fluoride into the product receiver. The traps are cooled to −40°C., and the trap-scrubber column and product receiver evacuated. (The sodium fluoride scrubber is operated at ambient temperature.) The product-receiver valve is opened fully, which is followed by the slow opening of the reactor outlet valve to initiate the transfer of perchloryl fluoride into the product receiver. The bulk of the accompanying hydrogen fluoride vapor will condense in the cold traps, while the remainder will be absorbed by the sodium fluoride in the scrubber. When the reactor outlet valve is fully

*Harshaw Chemical Company, 1933 E. 97th St., Cleveland, Ohio 44100.

opened, the lower portion of the product receiver is cooled with liquid nitrogen and the valve closed in 1-2 minutes.*

The product yield may be established by pressure measurements on the receiver of known volume or by weighing. The purity of the product may be confirmed by infrared spectrophotometry[7] or mass spectrometry. The only significant impurity found was hydrogen fluoride in trace amounts. This may be determined acidimetrically by passing a measured quantity of the product through water.

Properties

Perchloryl fluoride is thermally stable up to $500°C$. and very resistant to hydrolysis. It is a colorless gas in ordinary conditions with b.p., $-46.7°C$., and m.p., $-147.8°C$. It is a powerful oxidant at elevated temperatures. It exhibits selective fluorination properties and has been used as a perchlorylation reagent for introducing the ClO_3 group on carbon in organic compounds. It is moderately toxic (maximum allowable concentration, 3 p.p.m.[8]). A comprehensive review of the production, physical properties, and reactions of perchloryl fluoride is available.[9]

References

1. H. Bode and E. Klesper, *Z. Anorg. Allgem. Chem.*, **266**, 275 (1951).
2. A. Englebrecht and H. Atzwanger, *Monatsh. Chem.*, **83**, 1087 (1952); *J. Inorg. Nucl. Chem.*, **2**, 348 (1956).
3. G. Barth-Wehrenalp, *J. Inorg. Nucl. Chem.*, **2**, 266 (1956); U.S. patent 2,942,948 (June 28, 1960); W. A. LaLande, Jr., U.S. patent 2,982,617 (May 2, 1961).
4. G. Barth-Wehrenalp and H. C. Mandell, Jr., U.S. patent 2,942,949 (June 28, 1960).
5. C. A. Wamser, W. B. Fox, D. Gould, and B. Sukornick, *Inorg. Chem.*, **7**, 1933 (1968).
6. E. H. Appleman and M. H. Studier, *J. Am. Chem. Soc.*, **91**, 4561 (1969).
7. A. Englebrecht and H. Atzwanger, *J. Inorg. Nucl. Chem.*, **2**, 354 (1956).
8. "Handling and Storage of Liquid Propellants," Chap. 19, Chemical Propulsion Information Agency, Silver Spring, Md., Jan. 1963.
9. V. M. Khutoretskii, L. V. Okhlobystina, and A. A. Fainzil'berg, *Russ. Chem. Rev. (English Transl.)*, **36**(3), 145 (1967).

*This operation ensures the transfer of the bulk of the perchloryl fluoride from the reaction vessel and associated lines. Concurrently, some anhydrous hydrogen fluoride from the $-40°C$. traps will tend to transfer also (vapor pressure of hydrogen fluoride at $-40°C$. = *ca*. 50 mm.) but will be absorbed in the sodium fluoride scrubber.

7. CHLORODIFLUOROAMINE AND DIFLUORODIAZENE

Submitted by LEON M. ZABOROWSKI,* RONALD A. DE MARCO,*
and JEAN'NE M. SHREEVE*
Checked by MAX LUSTIG†

Chlorodifluoroamine has been prepared by reaction of difluoroamine with boron trichloride,[1] phosgene (carbonyl chloride),[2] or
hydrogen chloride;[2] treating a mixture of sodium azide and
sodium chloride with fluorine;[3] reaction of chlorine trifluoride
with ammonium fluoride;[4] reaction of chlorine with difluoroamine in the presence of potassium fluoride;[5] and photolysis of
tetrafluorohydrazine and sulfinyl chloride (thionyl chloride).[6]

Difluorodiazene has been prepared by the thermal decomposition of fluorine azide,[7] electrolysis of ammonium hydrogen
fluoride,[8] reaction of nitrogen trifluoride with mercury vapor in
an electric discharge,[9] dehydrofluorination of difluoroamine,[10]
treatment of a solution of 1,1-difluorourea with a concentrated
potassium hydroxide solution,[11] reaction of sodium azide with
fluorine,[12] decomposition of $N_2F_3Sb_2F_{11}$,[13] and reaction of
tetrafluorohydrazine with excess aluminum chloride at $-78°$.[14]
However, each of these methods suffers from one or more
disadvantages, including low[4,8,9,11,12,14] or erratic[2,3] yields,
tendency to explode,[3,4,7] use of a very hazardous reagent[1,2,5,10]
(difluoroamine is extremely shock sensitive as a solid), and
indirect method of preparation.[13]

The following are convenient methods for the preparation of
chlorodifluoroamine and difluorodiazene in reproducibly good
yields by the reaction of N,N-difluorohydroxylamine-O-sulfonyl
fluoride with sodium chloride and the photolysis of a tetrafluorohydrazine with bromine.

*University of Idaho, Moscow, Idaho 83843.
†Memphis State University, Memphis, Tenn. 38111.

■ *Caution. Care should be exercised in handling tetrafluoro-hydrazine, chlorodifluoroamine, and difluorodiazene because N-halogen compounds are known to exhibit explosive properties; safety shielding and gloves should be used. Any apparatus used should be clean and free of organic materials. Liquid nitrogen should be used for condensing reagents.*

In the following procedures a standard glass vacuum line with high-vacuum stopcocks (lubricated with Kel-F-90 grease*) is used. Because of the reactivity of many of the compounds with mercury, it is convenient to use a null-point pressure device, such as a Booth-Cromer[16] pressure gage or spiral gage. A mercury manometer covered with Kel-F-3 oil* can be used.

Procedure

A. CHLORODIFLUOROAMINE
(Nitrogen Chloride Difluoride)

$$2N_2F_4 + 2SO_3 \xrightarrow{h\nu} 2NF_2OSO_2F + N_2F_2$$

$$NF_2OSO_2F + NaCl \xrightarrow{CH_3CN} NF_2Cl + NaOSO_2F$$

N,N-Difluorohydroxylamine-O-sulfonyl fluoride is prepared by the photolysis of tetrafluorohydrazine and sulfur trioxide (55% yield)[17] or essentially quantitatively by the reaction of N_2F_4 and peroxodisulfuryl difluoride $(S_2O_6F_2)$.[18]

A 300-ml. Pyrex glass vessel fitted with a Teflon resin stopcock† and containing a Teflon resin-coated stirring bar is charged with excess reagent-grade sodium chloride (0.052 mole). After evacuation on the vacuum line, 3 ml. of dry acetonitrile and then N,N-difluorohydroxylamine-O-sulfonyl fluoride (0.010 mole) are distilled into the vessel, which is at $-195°C$. The mixture is warmed to room temperature and is stirred with a magnetic stirring device for 2 hours (behind a safety shield). The volatile compounds are removed under static vacuum from the reaction

---

*3M, Minnesota Mining and Manufacturing Company, St. Paul, Minn. 55119.
†Fischer and Porter Company, Warminster, Pa. 18974.

vessel, which is held at $-78°C$. (to retain acetonitrile), to a trap at $-195°C$. Then as the latter warms from $-195°C$., the material is separated by passing through traps at -135 and $-195°C$. The first trap will contain acetonitrile and small amounts of unreacted starting material, whereas the trap at $-195°C$. will contain pure chlorodifluoroamine (0.0094 mole, >90%).[19] Chlorodifluoroamine passes the trap at $-135°C$. slowly under good vacuum. Although chlorodifluoroamine can be stored for long periods in Pyrex glass at $-195°C$., for reasons of safety it is suggested that only small amounts (<0.01 mole) be retained.

B. DIFLUORODIAZENE
(Dinitrogen Difluoride)

$$N_2F_4 \xrightarrow[Br_2]{h\nu} N_2F_2 + \text{complex products}$$

Reagent-grade bromine is used without further purification. It can be stored under static vacuum for long periods at room temperature in an ordinary Pyrex glass tube equipped with a Teflon resin* stopcock. Tetrafluorohydrazine† is used without further purification.

Photolysis is carried out in an 850-ml. Pyrex glass vessel equipped with a water-cooled quartz probe. The ultraviolet light source is a 450-watt lamp‡ with a Vycor filter‡. To reduce the dangers from a possible explosion or eye damage from ultraviolet radiation, the reaction vessel shown in Fig. 5 is contained in a wooden box.

The photolysis bulb is connected to the vacuum line by a 10/30 S.T. joint and is evacuated. Bromine (0.004 mole) and tetrafluorohydrazine (0.002 mole) are condensed into the cold finger, A in Fig. 5, at $-195°C$. The mixture expands into the bulb as it warms to room temperature. The lamp is turned on only after the Vycor

---

*Fischer and Porter Company, Warminster, Pa. 18974.

†Air Products and Chemicals, Box 538, Allentown, Pa. 18100.

‡Hanovia L-679A36 and filter 7910, Hanovia Lamp Division, Engelhard Industries, 100 Chestnut St., Newark, N.J. 07100.

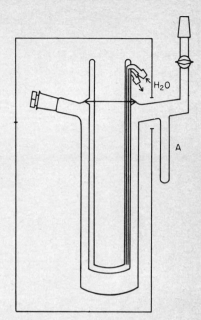

Fig. 5. Pyrex glass photolysis vessel with water-cooled quartz probe.

filter is in place, the cold tap water is passing through the water jacket, and the reagents are at room temperature. Photolysis time for an 850-ml. bulb is about 90 minutes.

After the photolysis is completed, the reaction mixture is transferred under dynamic vacuum to a trap at $-195°$C. The small amount of nitrogen formed in the reaction is expelled by the pumping system. The mixture is allowed to warm slowly to room temperature (an empty Dewar flask cooled to $-195°$C. with liquid nitrogen is convenient), and a trap-to-trap separation is performed by using traps at -140 and $-195°$C. The first trap contains N_2O_4 and Br_2. The photolysis vessel contains a white solid, probably $(NO)_2SiF_6$.

Difluorodiazene contaminated with SiF_4 and NF_3 is held at $-195°$C. This mixture is passed through a sodium fluoride trap to remove SiF_4 by the formation of Na_2SiF_6. Difluorodiazene may be separated from nitrogen trifluoride by gas chromatography with the use of a 25-ft. X 0.25-in. aluminum or copper column packed with 20% FC-43* on acid-washed Chromasorb P. A helium flow rate of 0.5 cc./second is used, and the column is held at $-63°C$. Nitrogen trifluoride, *trans*-difluorodiazene, and *cis*-difluorodiazene elute in that order. The yield is 70% difluorodiazene (53% trans). With a 5-l. bulb, and with 0.015 mole bromine and 0.009 mole tetrafluorohydrazine, the same yield results after 90 minutes of photolysis.

Although difluorodiazene can be stored for long periods in Pyrex glass at $-195°C$. or in metal at room temperature, for reasons of safety it is suggested that only small amounts (<0.01 mole) be retained.

Properties

Chlorodifluoroamine is a white solid at $-195°C$. and a colorless liquid at $-184°C$. Its normal boiling point is $-67°C$. The vapor-pressure curve is given by the equation $\log P_{mm.} = -950/T + 7.478$. The infrared spectrum consists of the following peaks: 1853(w), 1755(w), 1695(w), 1372(w), 926(s), 855(s), 746(m), 694(s) cm.$^{-1}$.[1] The ^{19}F n.m.r. shows a broad triplet centered at -141.5 p.p.m. relative to an internal reference of CCl_3F.

cis-Difluorodiazene is a colorless liquid at $-195°C$. with a boiling point of $-105.7°C$. The vapor-pressure curve follows the equation $\log P_{mm.} = 803.0/T + 7.675$. The infrared spectrum consists of the following peaks: 1538(w), 1513(w), 954(s), 904(s), 883(s), 738(vs) cm.$^{-1}$.[20] The ^{19}F n.m.r. shows a broad triplet centered at -136.1 p.p.m., relative to an internal reference of CCl_3F.

trans-Difluorodiazene is a white solid, melting at $-172°C$. to a liquid that boils at $-111.4°C$. The vapor-pressure curve is given by the equation $\log P_{mm.} = -742/T + 7.470$. The infrared spectrum is a strong band at 995 cm.$^{-1}$.[20] The ^{19}F n.m.r. spectrum is a broad triplet centered at -94.4 p.p.m. relative to an internal reference of CCl_3F.

*3M, Minnesota Mining and Manufacturing Company, St. Paul, Minn. 55119. The checker reports substantially identical results with perfluorotri-*tert*-butylamine on Chromasorb P.

References

1. R. C. Petry, *J. Am. Chem. Soc.*, **82**, 2400 (1960).
2. E. A. Lawton and J. Q. Weber, *ibid.*, **85**, 3595 (1963).
3. T. A. Austin and R. W. Mason, *Inorg. Chem.*, **2**, 646 (1963).
4. D. M. Gardner, W. W. Knipe, and C. J. Mackley, *ibid.*, **2**, 413 (1963).
5. W. C. Firth, Jr., *ibid.*, **4**, 254 (1965).
6. L. M. Zaborowski, K. E. Pullen, and J. M. Shreeve, *ibid.*, **8**, 2005 (1969).
7. J. F. Haller, doctoral dissertation, Cornell University, Ithaca, N.Y., 1942.
8. C. B. Colburn, F. A. Johnson, A. Kennedy, K. McCallum, L. C. Metzger, and C. O. Parker, *J. Am. Chem. Soc.*, **81**, 6397 (1959).
9. J. W. Frazer, *J. Inorg. Nucl. Chem.*, **11**, 166 (1959).
10. E. A. Lawton, D. Pilipovich, and R. D. Wilson, *Inorg. Chem.*, **4**, 118 (1965).
11. F. A. Johnson, *ibid.*, **5**, 149 (1966).
12. H. W. Roesky, O. Glemser, and D. Bormann, *Chem. Ber.*, **99**, 1589 (1966).
13. J. K. Ruff, *Inorg. Chem.*, **5**, 1791 (1966).
14. G. L. Hurst and S. I. Khayat, *J. Am. Chem. Soc.*, **87**, 1620 (1965).
15. R. C. Petry, *ibid.*, **89**, 4600 (1967).
16. S. Cromer, "The Electronic Pressure Transmitter and Self Balancing Relay," *U.S. At. Energy Comm.*, MDDC-803, declassified Mar. 20, 1947.
17. G. W. Fraser, J. M. Shreeve, M. Lustig, and C. L. Bumgardner, *Inorganic Syntheses*, **12**, 299 (1970).
18. M. Lustig and G. H. Cady, *Inorg. Chem.*, **2**, 388 (1963).
19. J. K. Ruff, *ibid.*, **4**, 1788 (1965).
20. C. B. Colburn, *Advan. Fluorine Chem.*, **3**, 104 (1963).

8. DIOXYGENYL SALTS

$$O_2 + \frac{1}{2}F_2 + AsF_5 \longrightarrow O_2^+AsF_6^-$$

$$O_2 + \frac{1}{2}F_2 + SbF_5 \longrightarrow O_2^+SbF_6^-$$

Submitted by JACOB SHAMIR* and JEHUDA BINENBOYM*†
Checked by J. G. MALM‡ and C. W. WILLIAMS‡

The few dioxygenyl salts described in the literature require the preparation of some highly unstable fluorinating agents, such as

*Department of Inorganic and Analytical Chemistry, The Hebrew University, Jerusalem, Israel.
†Present Address: Department of Inorganic Chemistry, Soreq Nuclear Research Center, Yavne, Israel.
‡Chemistry Division, Argonne National Laboratory, Argonne, Ill. 60439.

platinum(VI) fluoride[1] or O_2F_2.[2,3] The synthesis of either of these requires special equipment and is rather difficult to handle. The platinum(VI) fluoride is very easily reduced, and the O_2F_2 is stable only at low temperatures.

A simple synthesis of these salts became available with the recognition that oxygen, fluorine, and arsenic or antimony-(V) fluoride, at pressures near atmospheric, react to form the appropriate dioxygenyl salts when exposed to ordinary sunlight filtered by Pyrex glass.[4] Thus, the atmospheric-pressure, photo-chemical synthesis is the safest and most convenient one to date.

Procedure

A 500-ml. Pyrex bulb, equipped with a male joint, is connected through a female joint to a glass, vacuum stopcock, and this, in turn, is connected with a proper joint to a vacuum line. The stopcock and the joints are lubricated with Kel-F grease. It is best to use an all-metal, vacuum line for the handling of fluorine and its reactive compounds. Such lines have been described in this series.[5] The reaction vessel should be cleaned of all organic material, dried thoroughly by flaming or baking under vacuum, and treated once or twice with fluorine at about 100-mm. pressure in sunlight for about an hour.

■ *Caution. Fluorine must be handled with great care and with special apparatus. See descriptions of apparatus and precautions for such handling in other contributions to this series.*[5]

Arsenic(V) fluoride is introduced into the Pyrex bulb to a pressure of 250 mm.* The stopcock is closed and the bulb immersed in Dry Ice or liquid nitrogen. [Antimony(V) fluoride has only a few millimeters of vapor pressure at 25°C., but it can easily be distilled into the bulb and the quantity determined by

---

*A suitable quantity of material for the preparation in a Pyrex vessel is about 250-mm. pressure of each gas.

weighing.] Equal amounts of fluorine and oxygen are mixed in the metal line in sufficient amount to give the desired pressure when expanded into the glass bulb. The stopcock is then closed, and the vessel is warmed to 25°C., detached from the line, and placed in daylight for direct exposure. The reaction takes place in a very short time. On a bright day the reaction is completed in a matter of minutes. An initial mixture of 1:1:1 of $O_2/F_2/AsF_5$ (SbF_5) will leave some unreacted fluorine, which is removed by pumping. The solid formed is decomposed by moisture and must be handled in an inert-atmosphere dry-box. The yields are greater than 95%. *Anal.* Calcd. for $O_2^+SbF_6^-$: Sb, 45.47; F, 42.58. Found: Sb, 48.11; F, 44.58.

Properties

Dioxygenyl hexafluoroarsenate(V) or antimonate(V) is a white solid.[2,4] It can be stored in predried nickel, glass, Kel-F, or Teflon vessels. It hydrolyzes in water or moist air, which produces equal amounts of oxygen and ozone ($2O_2^+AsF_6^- + H_2O \rightarrow O_2 + O_3 + 2HAsF_6$). The compounds can be easily identified by their Raman spectra, which show, in addition to the three MF_6^- bands (ν_1, ν_2, and ν_5), a strong line at 1858–1862 cm.$^{-1}$.[6] X-ray data[2,4] indicate that both compounds are cubic; a_0 = 8.10 A. for dioxygenyl hexafluoroarsenate(V), and a_0 = 10.30 A. for dioxygenyl hexafluoroantimonate(V). The paramagnetic susceptibility has been determined[7] and obeys the equation:

$$\chi = \frac{0.309}{T - 0.07}$$

References

1. N. Bartlett and D. H. Lohmann, (a) *Proc. Chem. Soc.,* 1962, 115; (b) *J. Chem. Soc.,* 1962, 5253.
2. A. R. Young II, T. Hirata, and S. I. Morrow, *J. Am. Chem. Soc.,* 86, 20 (1964).

3. I. J. Solomon, R. I. Brabets, R. K. Uenishi, J. N. Keith, and J. M. McDonough, *Inorg. Chem.*, **3**, 457 (1964).
4. J. Shamir and J. Binenboym, *Inorg. Chim. Acta*, **2**, 37 (1968).
5. F. A. Hohorst and J. M. Shreeve, *Inorganic Syntheses*, **11**, 143 (1968).
6. J. Shamir, J. Binenboym, and H. H. Claassen, *J. Am. Chem. Soc.*, **90**, 6223 (1968).
7. A. Grill, M. Schieber, and J. Shamir, *Phys. Rev. Letters*, **25**, 747 (1970).

9. BIS(TRIFLUOROMETHYL) SULFOXIDE

$$CF_3 SCl + AgOCOCF_3 \longrightarrow CF_3 SOCOCF_3 + AgCl$$

$$CF_3 SOCOCF_3 \xrightarrow{h\nu} CF_3 SCF_3 + CO_2$$

$$CF_3 SCF_3 + 2ClF \xrightarrow[\text{10 hr.}]{-78°C. \text{ to } 25°C.} CF_3 SF_2 CF_3 + Cl_2$$

$$CF_3 SF_2 CF_3 + HCl \xrightarrow{\text{Pyrex}} [CF_3 SCl_2 CF_3] + 2HF$$

$$4HF + SiO_2 \longrightarrow 2H_2 O + SiF_4$$

$$[CF_3 SCl_2 CF_3] + H_2 O \longrightarrow CF_3 S(O)CF_3 + 2HCl$$

Submitted by DENNIS T. SAUER* and JEAN'NE M. SHREEVE*
Checked by MAX LUSTIG†

Bis(trifluoromethyl) sulfoxide has been prepared[1] previously by the direct fluorination of bis(trifluoromethyl) sulfide at −78°C. in hexafluoroethane followed by hydrolysis of the bis(trifluoromethyl)sulfur difluoride (difluorobis(trifluoromethyl)sulfur). This method suffers because elemental fluorine must be used, and the yields are low.

Oxidation of bis(trifluoromethyl) sulfide with commercially obtainable chlorine monofluoride in the absence of solvent yields bis(trifluoromethyl)sulfur difluoride in >90% yield.[2,3] Pure bis-(trifluoromethyl)sulfur difluoride is resistant to hydrolysis and is stable in Pyrex glass at 25°C. for extended periods of time. Reaction of bis(trifluoromethyl)sulfur difluoride with anhydrous

*University of Idaho, Moscow, Idaho 83843.
†Memphis State University, Memphis, Tenn. 38111.

hydrogen chloride in a clean Pyrex bulb results in the formation of bis(trifluoromethyl) sulfoxide in good yield. This preparative method has been extended and results in the preparation of $CF_3S(O)C_2F_5$, $CF_3S(O)C_3F_7$, and $C_2F_5S(O)C_2F_5$.[2,3]

Bis(trifluoromethyl) sulfide was prepared by the photolysis of S-trifluoromethyl trifluoromonothioperoxyacetate (trifluoroacetic trifluorosulfenic anhydride).[4] Other preparative methods[5,6] have been difficult to reproduce, or they produce the monosulfide in small yield. Oxidation of bis(trifluoromethyl) sulfide with chlorine monofluoride proceeds smoothly when the metal reactor containing the mixture is slowly warmed from -78 to $25°C$. over 10 hours. No cleavage products are formed, and the desired bis(trifluoromethyl)sulfur difluoride is isolated in $>90\%$ yield. The reaction of the sulfur difluoride with anhydrous hydrogen chloride to produce bis(trifluoromethyl) sulfoxide is presumed to proceed through the bis(trifluoromethyl)sulfur dichloride intermediate. Since hydrogen fluoride is produced when hydrogen chloride reacts with bis(trifluoromethyl)sulfur difluoride in Pyrex glass, water is formed, which results in hydrolysis of the bis(trifluoromethyl)sulfur dichloride intermediate. Attempts to isolate the sulfur dichloride intermediate by reaction of hydrogen chloride and bis(trifluoromethyl)sulfur difluoride in the presence of sodium fluoride in a nickel bomb resulted in the formation of bis(trifluoromethyl) sulfide and chlorine quantitatively.

A. S-TRIFLUOROMETHYL TRIFLUOROMONOTHIOPEROXYACETATE
(Trifluoroacetic Trifluorosulfenic Anhydride)

Procedure

Ten mmoles of trifluoromethanesulfenyl chloride,* CF_3SCl, is allowed to react with excess silver trifluoroacetate* at $25°C$. for

*Peninsular Chemical Products Company, 6801 E. 9 Mike at Weiner, Warren, Mich. 48089.

10 minutes in a 1-l. Pyrex vessel to produce S-trifluoromethyl trifluoromonothioperoxyacetate, $CF_3SOCOCF_3$. The reaction is quantitative. The $CF_3SOCOCF_3$ may be freed from trace amounts of trifluoromethanesulfenyl chloride, CF_3SCl, by passage through a $-78°C$. Dry Ice–acetone bath, which retains the pure CF_3-$SOCOCF_3$.

Properties

S-Trifluoromethyl trifluoromonothioperoxyacetate is a colorless liquid at $25°C$. The ^{19}F n.m.r. resonances occur at 47.3 p.p.m. (CF_3S) and 76.5 p.p.m. $(CF_3C(O)O)$ relative to CCl_3F. No coupling is observed between the trifluoromethyl groups.[4] The infrared spectrum consists of bands at 1835(m), 1805(w,sh), 1317(w), 1246(m-s), 1202(vs), 1190(s,sh), 1120(m-s), 1069(s), 835(w), 765(w-m), and 720(w) cm^{-1}.

B. BIS(TRIFLUOROMETHYL) SULFIDE

■ *Caution. The volatile reactants and products are toxic and contact with these reagents should be avoided.*

Procedure

Ten mmoles of S-trifluoromethyl trifluoromonothioperoxy-acetate are photolyzed for $\frac{3}{4}$ hour through Pyrex glass with a Hanovia Utility ultraviolet quartz lamp (140 watts), which produces bis(trifluoromethyl) sulfide and carbon dioxide quantitatively. Pure bis(trifluoromethyl) sulfide is retained in a $-120°C$. slush bath (diethyl ether) while carbon dioxide slowly sublimes into a $-183°C$. bath during trap-to-trap distillation.

Properties

Bis(trifluoromethyl) sulfide exists as a colorless gas at room temperature and condenses to a colorless liquid. The vapor pressure of bis(trifluoromethyl) sulfide is given by the equation $\log P_{mm.} = 7.82 - 1239.1/T$, from which the b.p. is calculated as $-22.2°C.$[5] The ^{19}F n.m.r. spectrum consists of a single resonance at 38.6 p.p.m. relative to CCl_3F. The infrared spectrum contains bands at 1220(s), 1198(vs), 1160(s), 1078(vs), 758(m), and 475(w) cm^{-1}.

C. BIS(TRIFLUOROMETHYL)SULFUR DIFLUORIDE

■ *Caution. Chlorine monofluoride is toxic and exceedingly damaging to the skin.*

Procedure

Reaction of bis(trifluoromethyl) sulfide with chlorine monofluoride* is carried out in a 75-ml. stainless-steel Hoke bomb. The bomb is evacuated, and in a typical preparation, 10 mmoles of bis(trifluoromethyl) sulfide and 22 mmoles of chlorine monofluoride are added at $-183°C$. The vessel is warmed to $-78°C$. and allowed to warm slowly to $25°C$. over a 10-hour period. The product mixture is first separated by fractional condensation. The bis(trifluoromethyl)sulfur difluoride ($CF_3SF_2CF_3$) is retained in a $-98°C$. slush bath, while any unreacted CF_3SCF_3, ClF, and Cl_2 pass into a $-183°C$. bath. The $CF_3SF_2CF_3$ may be purified further by gas chromatography utilizing a 17-ft., 20% Kel-F oil† on Chromasorb P column. Final purification gives $CF_3SF_2CF_3$ in >90% yield based on the amount of monosulfide used.

Properties

At $25°C$. bis(trifluoromethyl)sulfur difluoride is a colorless gas which condenses, on cooling, to a colorless liquid. A boiling point of

*Ozark-Mahoning Company, 1870 S. Boulder Ave., Tulsa, Okla. 74119.
†3M, Minnesota Mining and Manufacturing Company, St. Paul, Minn. 55119.

21°C. is calculated from the Clausius-Clapeyron equation $\log P_{mm.}$ = 8.00 − 1507/T. The ^{19}F n.m.r. resonances at 58.0 (CF$_3$) and 14.2 p.p.m. (SF$_2$) relative to CCl$_3$F integrate to the proper 6:2 ratio with $J_{SF_2-CF_3}$ = 19.5 Hz. The infrared spectrum contains bands at 1281(vs), 1260(s), 1215(m–s), 1144(m), 1081(vs), 766(m), 677(s), and 507(m) cm^{-1}.

D. BIS(TRIFLUOROMETHYL) SULFOXIDE

Procedure

Four mmoles of bis(trifluoromethyl)sulfur difluoride react with 16 mmoles of anhydrous hydrogen chloride* in a clean, 1-l. Pyrex vessel for 24 hours to give bis(trifluoromethyl) sulfoxide in 70% yield. The bis(trifluoromethyl) sulfoxide is purified by fractional condensation. The desired sulfoxide is retained in a −78°C. bath while unreacted hydrogen chloride and bis(trifluoromethyl)sulfur difluoride pass into a −183°C. bath. Further purification by gas chromatography, utilizing a 17-ft., 20% Kel-F oil on Chromasorb P column, enables isolation of pure CF$_3$S(O)CF$_3$. When the reaction is carried out in a metal bomb, no sulfoxide is formed. The products isolated were identified as CF$_3$SCF$_3$, chlorine, and unreacted hydrogen chloride.

Properties

Bis(trifluoromethyl) sulfoxide is a colorless liquid at 25°C. A normal boiling point of 37.3°C. is calculated from the Clausius-Clapeyron equation $\log P_{mm.}$ = 7.66 − 1483/T. Confirmatory spectral properties include a molecular ion in the mass spectrum (2.1%) and a single ^{19}F resonance at 64.5 p.p.m. relative to CCl$_3$F. The infrared spectrum contains bands at 1244(vs), 1191, 1187 (doublet, s), 1121(m–s), 1105(vs), 752(w), and 468(w) cm^{-1}.

*Matheson Gas Products, East Rutherford, N.J. 07073.

References

1. E. W. Lawless, *Inorg. Chem.*, 9, 2796 (1970).
2. D. T. Sauer and J. M. Shreeve, *Chem. Commun.*, 1970, 1679.
3. D. T. Sauer and J. M. Shreeve, *J. Fluorine Chem.*, 1, 1 (1971).
4. A. Haas and D. Y. Oh, *Chem. Ber.*, 102, 77 (1969).
5. G. A. R. Brandt, H. J. Emeléus, and R. N. Haszeldine, *J. Chem. Soc.*, 1952, 2198.
6. E. W. Lawless and L. D. Harman, *J. Inorg. Nucl. Chem.*, 31, 1542 (1969).

10. ALUMINUM TRIHYDRIDE-DIETHYL ETHERATE

(Etherated Alane)

$$3LiAlH_4 + AlCl_3 \xrightarrow{\text{diethyl ether}} 4[AlH_3 \cdot 0.3(C_2H_5)_2O] + 3LiCl$$

Submitted by D. L. SCHMIDT,* C. B. ROBERTS,* and P. F. REIGLER*
Checked by M. F. LEMANSKI, JR.,† and E. P. SCHRAM†

Methods for the preparation of aluminum trihydride-diethyl etherate, $AlH_3 \cdot 0.3[(C_2H_5)_2O]$,‡ have been published,[1,2] but the absence of complete experimental details makes duplication difficult. The following procedure is a modification of that reported by Finholt, Bond, and Schlesinger.[1] Problems inherent in previous methods, such as premature precipitation, decomposition of the alane, and lithium chloride contamination, are avoided.

Premature precipitation is controlled by maintaining a low temperature ($-5°$C.) in the reaction mixture. Purified reactants, as well as minimal exposure to light and higher temperatures,

*The Dow Chemical Company, Midland, Mich. 48640. This work was supported by ARPA and the Air Force under contracts AF 33(616)-6149 and AF 04(611)-7554.

†Ohio State University, Columbus, Ohio 43210.

‡Ebulliometric molecular weight determinations obtained in diethyl ether (concentration range, 0.25–1.0 *M*) indicates molecular weights only slightly higher than 30. This is consistent with studies reported by Wiberg and Uson.[3] It must be assumed that in solution, AlH_3 exists essentially as the monomeric form.

decrease the decomposition of aluminum trihydride. Lithium chloride contamination is avoided by a subsequent treatment of the solvated aluminum trihydride solution with sodium tetrahydroborate.

Procedure

All operations must be performed in the absence of water, oxygen, and other species reactive with the starting materials or products. The reactions are carried out in a nitrogen- or argon-filled dry-box, with the concentration of oxygen and water no higher than 3–4 p.p.m.* Commercially available diethyl ether, $LiAlH_4$, and $AlCl_3$ require further purification. Since small amounts of impurities cause decomposition† of the aluminum trihydride, all reagents must be anhydrous and contamination minimized. Glass apparatus should be washed with red fuming nitric acid or hydrofluoric acid and rinsed copiously with distilled water. The apparatus is dried at 110–120°C. before being placed in the dry-box.

Diethyl ether (analytical reagent) is purified by distillation, under either a nitrogen atmosphere or vacuum, from lithium tetrahydroaluminate (lithium aluminum hydride), $LiAlH_4$, and is then collected in the additional funnel. The ether may also be dried by passing it through a 40-cm. column of molecular sieves (type 13X, $\frac{1}{8}$-in. pellets, Linde‡). Because diethyl ether is highly flammable, its purification should be carried out in a hood. The $LiAlH_4$ is recrystallized by dissolving it in diethyl ether, filtering through a medium-porosity glass frit, and removing the solvent under reduced pressure. Aluminum chloride§ is obtained by

---

*Water can be determined by a Gilbarco sorption hygrometer model SHL-100. Oxygen can be monitored by means of an analytic system analyzer. Dry-boxes should be vented into a hood.

†Transition metals are especially harmful. See reference 4.

‡J. T. Baker Co., Phillipsburg, N.J.

§A solvent-free dry-box should be used to handle $AlCl_3$.

high-vacuum sublimation at 110–120°C. of reagent-grade material. At this temperature a sublimation of 50 g. requires about 24 hours. Passing the vapor through about 1–2 cm. of activated coconut charcoal (50–200 mesh*) increases the purity of $AlCl_3$ and the stability of the resulting aluminum trihydride. A convenient sublimation apparatus which can easily be taken through a dry-box vacuum interchange is shown in Fig. 6.

Chloride is removed from the ether solution of aluminum trihydride by the addition of sodium tetrahydroborate. The $LiBH_4$ remains in solution, while the sodium chloride precipitates and is removed by filtration. Best results are obtained when $NaBH_4$† is dried under vacuum at 60°C. for 8 hours and ground to 1 μm. or smaller. This last step is particularly important since both the reaction rate and efficiency of chloride removal are a function of $NaBH_4$ particle size.

In a dry-box, a magnetic stirring bar and 32.3 g. (0.242 mole) of aluminum chloride, $AlCl_3$, are placed in a 500-ml., round-bottomed flask fitted with a 35/25 ball joint. The capped flask is removed from the dry-box and quickly attached to an addition funnel containing ether under low nitrogen pressure ($\frac{1}{2}$–1 p.s.i.).

The 500-ml. flask and magnetically stirred contents are cooled in a Dry Ice–methylene chloride bath, which is followed by slow

*Fisher Scientific Co., Pittsburgh, Pa. 15219.
†Sodium tetrahydroaluminate ($NaAlH_4$) may be substituted.

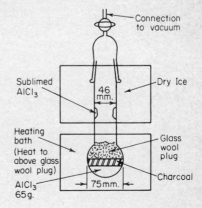

Fig. 6. Apparatus for purification of AlCl₃.

addition of 300 ml. of distilled ether from the addition funnel. After the ether has been cooled, the bath is removed. If the ensuing exothermic aluminum chloride solvation reaction causes bubbling of the ether, the flask is cooled again. After dissolution of the $AlCl_3$, the solution is allowed to warm to room temperature, tightly capped, and placed in a dry-box.

The $LiAlH_4$ solution is prepared by dissolving 28.2 g. (0.743 mole) of purified $LiAlH_4$ in about 750 ml. of distilled ether in a 1500-ml. Florence flask. Ether solutions of both $LiAlH_4$ and $AlCl_3$ are cooled to $-5°C$. by exposing them to a nitrogen purge which cools by evaporation.* Ether must be added during this period to keep the volume of the solutions reasonably constant.

The $AlCl_3$ solution is added slowly to the magnetically stirred, cold $LiAlH_4$ solution. After addition is complete, the reaction mixture is filtered by means of 3–5 p.s.i. nitrogen pressure through a medium-porosity glass frit into a 1500-ml. Florence flask containing 10 g. of $NaBH_4$ (particle size, 1 μm. or less) and a magnetic stirring bar. After being stirred for 3–4 minutes under a nitrogen purge, the solution is filtered again into a second 1500-ml. Florence flask, which removes excess $NaBH_4$ and precipitated sodium chloride.

The volume of clear aluminum trihydride solution is reduced to about 600 ml. by nitrogen-stream evaporation and then allowed to warm to room temperature. After 6 to 8 hours, the aluminum trihydride-diethyl etherate precipitates, which leaves the small amount of soluble $LiBH_4$ and aluminum trihydride-diethyl etherate in the ether solution. The product is filtered, washed twice with 100 ml. of anhydrous ether, and dried under high vacuum for 10 to 12 hours. A yield of approximately 30 g. (58%) is obtained. *Anal.* Calcd. for $AlH_3 \cdot 0.3[(C_2H_5)_2O]$ ‡: Al, 50.78; C, 28.21; H, 11.61; Hydrolytic H_2, 5.69. Found: Al, 50.60; C, 28.08; H, 11.62; Hydrolytic H_2, 5.68.

*The cold solution is necessary to prevent premature precipitation of the aluminum trihydride.
‡Calculated for $0.312[(C_2H_5)_2O]$.

Properties

The aluminum trihydride-diethyl etherate is a colorless solid which reacts violently with water or in a damp atmosphere. (■ *Caution. Sometimes this reaction is explosive!*) The material is unstable to prolonged exposure to light at ambient temperatures, but can be stored at $-10°C$. in a sealed container up to one year. The ratio of aluminum trihydride to ether $(C_2H_2)_2O$ in the solid varies between 0.29 and 0.33, depending upon the time under vacuum. The solubility of aluminum trihydride-diethyl etherate in diethyl ether is $0.2M$; it is very soluble in tetrahydrofuran.* Its infrared spectrum has broad bands in the

TABLE I X-Ray Powder Diffraction Pattern— 1443 Phase

d, A	I/I
12.0	100
4.62	17
3.89	17
2.88	17
2.35	4
2.31	4
1.54	2.7
1.495	3.3

*■ *Caution. Solid aluminum trihydride tetrahydrofuran often decomposes explosively when exposed to a high vacuum.*

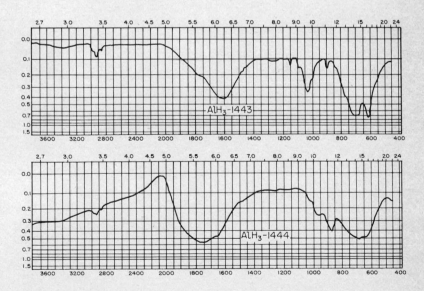

Fig. 7. Infrared spectra of etherate 1443 and 1444 of aluminum hydride. Top: Taken sometime in 1961. Bottom: Example of other phase.

regions of 1650–1850 and 600–800 cm.$^{-1}$ (Fig. 7). It can be identified by its x-ray powder pattern given in Table I. The principal impurities are Li, Cl, and B; in a typical example, <0.1% Li, <0.01% Cl, and <0.02% B were found.

References

1. A. E. Finholt, A. C. Bond, and H. I. Schlesinger, *J. Am. Chem. Soc.,* **69,** 1199 (1947).
2. E. Wiberg and M. Schmidt, *Z. Naturforsch.,* **66,** 333 (1951).
3. E. Wiberg and R. Uson, *Rev. Acad. Cienc. Exact., Fis. Quim. Nat. Zaragoza.,* **10**(2), 35 (1955).
4. D. L. Schmidt and R. Hellmann, British patent 1,122,359 (1967); U.S. patent 3,462,288 (1969).
5. W. W. Wendlant and R. Dunham, *Anal. Chim. Acta,* **19,** 505 (1958).

11. TRIMETHYLAMMONIUM TETRAPHENYLBORATE

(*A Ready Source of Triphenylborane*[1])

$$4LiC_6H_5 + BCl_3 \cdot ether \longrightarrow LiB(C_6H_5)_4 + ether + 3LiCl$$
$$LiB(C_6H_5)_4 + (CH_3)_3NHCl \longrightarrow (CH_3)_3NHB(C_6H_5)_4 + LiCl$$

Submitted by K. E. REYNARD,* R. E. SHERMAN,* HAMPTON D. SMITH, JR.,* and L. F. HOHNSTEDT*
Checked by G. GASSENHEIMER† and T. WARTIK†

Wittig and co-workers[2] have shown that triphenylborane can be readily synthesized in high yield by simple thermal decomposition of trimethylammonium tetraphenylborate. They obtained the latter compound by collecting the precipitate which was formed on mixing aqueous solutions of trimethylammonium chloride and lithium tetraphenylborate. The lithium salt was isolated from the products of the reaction of boron trifluoride-etherate with

*St. Louis University, St. Louis, Mo. 63103.
†Pennsylvania State University, University Park, Pa. 16802.

phenyllithium analogous to the above equation. A modification of the Wittig procedure[2] is described. The procedure requires about 4 hours, and the product obtained is a stable salt which does not demand the special handling and storage required for triphenyl-borane. As a dry solid, samples may be stored in screw-cap bottles exposed to the atmosphere, and the compound appears to be stable indefinitely at room temperature.

Samples of triphenylborane are obtained readily by thermal decomposition of trimethylammonium tetraphenylborate prepared by the method described. Preparation of triphenylborane by this method eliminates the tedious and somewhat hazardous treatment of the reactive residues produced during the formation of the triphenylborane by the Grignard method.[3]

Procedure

Phenyllithium* (0.586 mole in 296 ml.) in a 7:3 volume mixture of benzene and ether is placed in a reaction flask fitted with a pressure-equalizing dropping funnel containing boron trichloride in excess ether (54.8 ml. of solution containing approximately 0.143 mole BCl_3) and with a reflux condenser equipped to provide a dry nitrogen atmosphere. The solution of boron trichloride-etherate is prepared with vacuum-line procedures similar to those reported for the preparation of boron trifluoride-etherate,[4] but measured amounts of excess ether and boron trichloride are mixed. The resulting solution is allowed to warm to room temperature and is used without further purification. Boron trichloride-etherate is added at a rate sufficient to maintain gentle reflux, which is prolonged by external heating for 30 minutes after addition. The reaction mixture is allowed to cool to room temperature, and the solvent is removed by distillation at atmospheric pressure. The residue retains some solvent and has an appearance similar to a clay mud. Addition of 500 ml. of water

*Alfa Inorganics, Ventron Corporation, P.O. Box 159, Beverly, Mass. 01915.

produces no evidence of reaction, and the residue dissolves upon vigorous stirring. A less-dense liquid phase separates from the mixture and is discarded. A solution of trimethylammonium chloride (16.0 g. in 100 ml. of water) is added slowly to the stirred aqueous phase. Immediately, trimethylammonium tetraphenylborate precipitates as a siltlike, white solid. It is recovered by filtration. The solid is washed with water and dried in a vacuum desiccator over phosphorus pentoxide. The product, obtained in 72% yield, melts over a range of several degrees around 160°C.[5,*] and gives good yields of triphenylborane. It has an infrared spectrum indistinguishable from that of the white solid obtained by mixing aqueous solutions of trimethylammonium chloride and sodium tetraphenylborate. A purified sample may be prepared by dissolving some of the crude product in dichloromethane, reducing the volume of solvent by evaporation under reduced pressure, and drying the recovered solid in a vacuum desiccator until a constant melting point, 165–166°C.,[5] is obtained. *Anal.* (purified sample) Calcd. for $C_{27}H_{30}BN$: C, 85.50; H, 7.96; B, 2.85; N, 3.69. Found: C, 84.86; H, 8.01; B, 2.87; N, 3.88 (boron analyzed by atomic absorption spectrophotometry with tetraethylammonium tetrahydroborate as a standard).

The procedure gives satisfactory results on reduced scales, e.g., one-fourth or one-sixth the quantities given above. In addition, a saturated solution of boron trichloride in ether may be used in place of the solution described. A saturated solution may be prepared by allowing boron trichloride gas to bubble into ether at room temperature until excess boron trichloride-etherate settles out of the liquid phase.

Properties

Trimethylammonium tetraphenylborate is a white, crystalline solid. It is very soluble in acetone and moderately soluble in dichloromethane. It is difficult to dry at room temperature.

*Wendlant and Dunham observed that the melting point of the samples depended on the rate of heating.

References

1. Taken in part from the B.S. thesis of Robert E. Sherman, St. Louis University, St. Louis, Mo., 1967
2. G. Wittig and P. Raff, *Ann. Chem.*, 573, 195 (1951).
3. E. Krause and R. Nitsche, *Chem. Ber.*, 55B, 1261 (1922).
4. H. Ramser and E. Wiberg, *ibid.*, 63, 1136 (1930).

12. TETRAKIS(ACETATO)DI-μ-AMIDO-DIBORANE

$$2(BClNH)_3 + 12CH_3COOH \longrightarrow 3[(CH_3COO)_2BNH_2]_2 + 6HCl$$

Submitted by D. T. HAWORTH*
Checked by L. A. MELCHER† and K. NIEDENZU†

The procedure described represents the conversion, through a change in the coordination number of boron and nitrogen, of a six-membered boron nitrogen ring to a four-membered boron nitrogen ring.[1] The absence of other reaction products in the direct union simplifies the purification problem.

Procedure

A 250-ml., three-necked flask is fitted with a reflux condenser topped with a calcium chloride drying tube, a dropping funnel, and a nitrogen gas inlet tube. Under a blanket of nitrogen, the reaction flask is charged with 5 g. (0.027 mole) of 2,4,6-trichloroborazine‡,[2-4] and the dropping funnel with 75 ml. of glacial acetic acid. The gas inlet tube is removed, a magnetic stirring bar is added to the flask, and this neck of the flask is stoppered. While its contents are being stirred, the glacial acetic

*Marquette University, Milwaukee, Wis. 53233.
†University of Kentucky, Lexington, Ky. 40506.
‡U.S. Borax Research Corp., Anaheim, Calif.

acid is added to the flask over a 15-minute period. The solution is heated to gentle reflux for 30 minutes, which is followed by filtration while still warm through a coarse filter into a 500-ml. flask. The clear filtrate is cooled to room temperature, and the excess solvent is removed by vacuum distillation on a rotary evaporator. The resulting white crystals (7.2 g.) are washed with small portions of ether, filtered, and dried under vacuum overnight. The yield of the product is 61.5%; m.p. 140–142°C. If an oily paste is obtained on removal of the solvent, this paste should be washed several times with ether, filtered through a frit, and dried in vacuum.

Properties

Tetrakis(acetato)di-μ-amido-diborane is a white crystalline compound which is not too sensitive to moisture. It can be stored in a nitrogen atmosphere in a refrigerator for long periods without decomposition. It is sparingly soluble in most organic solvents and slowly dissolves in glacial acetic acid and acetic anhydride. Monoclinic and triclinic crystalline forms were obtained by recrystallization from acetic anhydride and glacial acetic acid, respectively.[1] The infrared spectrum recorded (Beckman i.r.-12) by the KBr pellet technique contains major absorption bands (at frequencies cm.$^{-1}$) 3280(s), 3230(s), 3100(s), 1740(w,sh), 1715(m), 1430(sh), 1390(vs), 1300(m), 1240(m), 1125(m), 1050(m), 1035(m), 900(w,sh), 775(w), and 700(w). The ^{1}H n.m.r. spectrum of the compound in $CDCl_3$ has a broad absorption at −4.9 p.p.m. (NH_2) and a singlet at −2.1 p.p.m. (CH_3). TMS is used as an internal standard.

References

1. G. L. Brennan, G. H. Dahl, and R. Schaeffer, *J. Am. Chem. Soc.*, **82**, 6248 (1960).
2. K. Niedenzu and J. W. Dawson, *Inorganic Syntheses*, **10**, 142 (1967).
3. E. F. Rothgery and L. F. Hohnstedt, *Inorg. Chem.*, **6**, 1065 (1967).
4. D. T. Haworth, *Inorganic Syntheses*, **13**, 41 (1972).

Chapter Three

TRANSITION-METAL COMPLEXES

13. A GENERAL NONAQUEOUS PREPARATION OF COBALT(III) AND NICKEL(II) DIAMINE AND TRIAMINE COMPLEXES

Submitted by KARL H. PEARSON,* WILLIAM R. HOWELL, JR.,†
PAUL E. REINBOLD,† and STANLEY KIRSCHNER‡
Checked by J. W. HART, JR.,§ J. KIM,§ R. PARKER,§
S. YASSINZADEH,§ and C. DAVID SCHMULBACH§

Previous syntheses of cobalt(III) and nickel(II) diamine and triamine complexes that have been carried out in aqueous solution require several hours, and can give products with varying quantities of water of hydration, depending on the drying techniques employed. For example, Jørgensen[1] prepared $[Co(en)_3]Cl_3$ by heating $[Co(NH_3)_5Cl]Cl_2$ with ethylenediamine (which involves a separate synthesis of the pentaammine complex), and Grossman and Schuck,[2] Werner,[3] Mann,[4] Work,[5] Jenkins and Monk,[6] State,[7] and Schlessinger[8] all used aqueous techniques, with air, oxygen, or peroxide oxidations for the cobalt(III) complexes. By

*Cleveland State University, Cleveland, Ohio 44115.
†Texas A and M University, College Station, Tex. 77843.
‡Wayne State University, Detroit, Mich. 48202.
§Southern Illinois University, Carbondale, Ill. 62901.

using nonaqueous techniques and different oxidants, the authors have found it possible to synthesize these complexes in a relatively short time (*ca.* 15–45 minutes), in relatively high yield (*ca.* 85–95%), with a very high degree of purity (see Tables I and II), and with no solvation problems.

For the cobalt(III) complexes this method offers the additional advantage of rapid, controlled oxidation of the cobalt(II) inter-mediate complex to the cobalt(III) complex through the direct use of the elemental halogen corresponding to the anion of the desired salt, which eliminates the long air oxidation and the possible undesirable side reactions.

A. COBALT COMPLEXES

Procedure

1. Cobalt(III) Bromides

$$CoCO_3 + 2HBr \longrightarrow CoBr_2 + H_2O + CO_2$$
$$2CoBr_2 + 6(amine) + Br_2 \xrightarrow{C_2H_5OH} 2[Co(amine)_3]Br_3$$

Reagent-grade cobalt(II) carbonate (2.38 g., 0.02 mole) is dissolved in 6.7 ml. of concentrated hydrobromic acid in a 1-l., three-necked, round-bottomed flask on a heating mantle. The flask is fitted with a dropping funnel and reflux condenser, and its contents are stirred with a magnetic stirrer. After complete dissolution of the salt, 500 ml. of absolute ethanol is added, and the mixture is allowed to reflux for 10–15 minutes. To the reflux-ing solution the desired redistilled, reagent-grade amine is added dropwise. Table I shows the volume of the amine to be used in all preparations of the cobalt complexes. Then 1.6 g. (0.52 ml.) of reagent-grade bromine, dissolved in 100 ml. of absolute ethanol, is added dropwise (over 20–30 minutes) to the refluxing solution. After the addition of the bromine is complete, the hot solution is

TABLE I Properties, Yields, and Analyses of the Cobalt(III) Complexes

Complex	Amine, ml.	%Yield	Concn. = 0.01 M λmax.	εmax.	Concn. = 0.005 M λmax.	εmax.	Analyses %C	%H	%N	%Br	%I
[Co(en)$_3$]Br$_3$*	5.5	94.4	466	68.7	466	64.6	Calcd.: 15.05	5.05	17.55	50.05	
							Found: 15.09	5.01	17.44	49.89	
[Co(en)$_3$]I$_3$	5.5	88.0	466	89.2	465	91.0	Calcd.: 11.63	3.90	13.56		61.41
							Found: 11.75	4.10	13.56		61.66
[Co(pn)$_3$]Br$_3$	6.8	86.0	466	104	466	108	Calcd.: 20.75	5.80	16.13	46.01	
							Found: 20.81	5.89	15.87	45.60	
[Co(pn)$_3$]I$_3$	6.8	86.0	466	92.5	466	95.4	Calcd.: 16.33	4.57	12.69		57.51
							Found: 16.29	4.78	12.50		57.76
[Co(dien)$_2$]I$_3$	6.4	82.0	463	99.5	463	101	Calcd.: 14.87	4.06	13.01		58.93
							Found: 14.95	4.41	13.08		59.04
[Co(chxn)$_3$]Br$_3$	15.0	50.0	474	101	471	90.2	Calcd.: 33.71	6.60	13.10	37.37	
							Found: 33.57	7.14	12.87	37.10	
[Co(chxn)$_3$]I$_3$	15.0	86.0	473†	103†	472‡	107‡	Calcd.: 27.64	5.41	10.74		48.66
							Found: 27.53	5.56	10.93		48.93

*Abbreviations used:

en   ethylenediamine(1,2-ethanediamine)   dien   diethylenetriamine(N-(2-aminoethyl)-1,2-ethanediamine)
pn   1,2-propylenediamine(1,2-propanediamine)   chxn   trans-1,2-cyclohexanediamine

†Because of the solubility limits, the concentration was 0.005 M.
‡Concentration is 0.0025 M.

59

filtered through a sintered-glass Büchner funnel. The crystals are then washed with absolute ethanol and then acetone, and dried at 50°C.

2. Cobalt(III) Iodides

$$CoCO_3 + 2HI \longrightarrow CoI_2 + H_2O + CO_2$$
$$2CoI_2 + 2x(\text{amine})^* + I_2 \xrightarrow[C_2H_5OH]{} 2[Co(\text{amine})_x]I_3$$

Reagent-grade cobalt(II) carbonate (2.38 g., 0.02 mole) is dissolved in 11 ml. of concentrated hydriodic acid in an 800-ml. beaker placed in a steam bath. Then 250 ml. of absolute ethanol is added. While the hot solution in the steam bath is being stirred with a magnetic stirrer, the desired redistilled, reagent-grade amine is added, which is followed immediately by the addition of 2.54 g. of reagent-grade iodine previously dissolved in 100 ml. of hot, absolute ethanol. Table I shows the volume of amine to be used. After standing with stirring for 10 minutes in the steam bath, the solution is cooled and filtered through a sintered-glass Büchner funnel. The collected product is washed with absolute ethanol and then acetone, and dried in an oven at 50°C.

Properties, Analyses, and Yields

All the cobalt(III) complexes mentioned in Table I are obtained as fine golden crystals. They are sparingly soluble in the light alcohols, and are essentially insoluble in most other organic solvents. They are all quite soluble in water, except for [Co(chxn)₃]-I₃,† which is sparingly soluble.

The spectral properties, microanalyses, and yields obtained for the various complexes are reported in Table I.

*For the bidentate chelating agents, x is 3; for the tridentate, x is 2.
†chxn = *trans*-1,2-cyclohexanediamine.

B. NICKEL COMPLEXES

Procedure

1. Nickel(II) Bromides

$$NiCO_3 + 2HBr \longrightarrow NiBr_2 + H_2O + CO_2$$
$$NiBr_2 + x(amine)* \xrightarrow[C_2H_5OH]{} [Ni(amine)_x]Br_2$$

Reagent-grade nickel(II) carbonate (1.19 g., 0.01 mole) is added slowly to 3.8 ml. (0.03 mole) of reagent-grade concentrated hydrobromic acid in a 250-ml. beaker. After complete dissolution of the first mixture, 100 ml. of reagent-grade absolute ethanol is added, and any spattering is rinsed down carefully. The solution is stirred magnetically. The specified volume given in Table II of the desired, redistilled, reagent-grade amine is added slowly, and the mixture continues to be stirred until precipitation is complete (absence of the blue color of the nickel ethanol complex in the supernate). In some instances it may be necessary to chill the solution to ensure complete precipitation. The propylenediamine and cyclohexanediamine complexes have been found to be soluble in the absolute ethanol. The precipitated product is filtered, washed with reagent-grade ethanol, and dried in an oven at 50°C.

2. Nickel(II) Iodides

$$NiCO_3 + 2HI \longrightarrow NiI_2 + H_2O + CO_2$$
$$NiI_2 + x(amine)* \xrightarrow[C_2H_5OH]{} [Ni(amine)_x]I_2$$

The direct preparation of the iodides may be accomplished in the same manner as the bromides, with the use of 5.5 ml. of reagent-grade hydriodic acid (47–50%) in place of the hydrobromic acid. The diethylenetriamine complex was found to be

*For the bidentate chelating agents, x is 3; for the tridentate, x is 2.

TABLE II Properties, Yields, and Analyses of the Nickel(II) Complexes

Complex	Amine, ml.	%Yield	Concn. 0.01 = M λmax. nm.	εmax.	Concn. = 0.005 M λmax. nm.	εmax.	Analyses %C	%H	%N	%Br	%I
[Ni(en)₃]Br₂*	4.2	98.1	543	6.2	545	6.0	Calcd.: 18.07	6.06	21.07	40.07	
			345	7.9	345	8.2	Found: 17.92	6.20	20.93	39.85	
[Ni(en)₃]I₂	4.2	93.9	544	6.6	544	6.4	Calcd.: 14.62	4.91	17.05		51.50
			345	8.6	345	8.4	Found: 14.57	5.21	16.90		51.57
[Ni(pn)₃]Br₂	6.8	84.2	544	6.7	543	7.0	Calcd.: 24.52	6.86	19.06	36.25	
			345	8.7	345	8.8	Found: 24.53	6.98	18.88	36.21	
[Ni(pn)₃]I₂	6.8	91.5	544	6.9	543	7.4	Calcd.: 20.23	5.65	15.72		47.46
			345	8.8	345	9.2	Found: 20.21	6.06	15.53		47.42
[Ni(dien)₂]Br₂	5.35	93.4	535	7.7	534	8.0	Calcd.: 22.62	6.06	19.78	37.62	
			347	10.2	346	10.2	Found: 22.66	6.02	20.00	37.79	
[Ni(dien)₂]I₂	5.35	93.5	534	7.7	535	8.2	Calcd.: 18.52	5.05	16.19		48.91
			346	10.2	346	10.4	Found: 18.52	5.06	15.85		49.08
[Ni(chxn)₃]Br₂	11.0	94.7	542†	7.4†	544‡	6.4‡	Calcd.: 38.55	7.55	14.98	28.50	
			343†	11.2	345‡	8.4‡	Found: 38.68	7.63	14.81	28.38	
[Ni(chxn)₃]I₂	11.0	97.3	...§	.....	...§	.....	Calcd.: 32.99	6.46	12.82		38.73
							Found: 32.83	6.45	12.82		38.96

*See Table I for abbreviations of ligand names.
†Because of solubility limits, the concentration was 0.005 M.
‡Concentration was 0.0025 M.
§Impossible to obtain the spectra because of the low solubility.

soluble in the absolute ethanol and required chilling before filtering.

Properties, Analyses, and Yields

All the nickel(II) complexes are obtained as fine violet crystals. They are insoluble in acetone, and all but the $[Ni(chxn)_3]I_2$* complex are water-soluble. The spectral properties, elemental analyses, and yields obtained for the various complexes are reported in Table II.

References

1. S. M. Jørgensen, *J. Prakt. Chem.,* [2], **39**, 8 (1889).
2. H. Grossman and B. Schuck, *Chem. Ber.,* **39**, 1889 (1906).
3. A. Werner, *ibid.,* **45**, 121 (1912).
4. F. G. Mann, *J. Chem. Soc.,* **1934**, 466.
5. J. B. Work, *Inorganic Syntheses,* **2**, 221 (1946).
6. I. L. Jenkins and C. B. Monk, *J. Chem. Soc.,* **1951**, 68.
7. H. M. State, *Inorganic Syntheses,* **6**, 200 (1960).
8. G. G. Schlessinger, "Inorganic Laboratory Preparations," pp. 189 ff., Chemical Publishing Company, Inc., New York, 1962.

14. DIANIONOBIS(ETHYLENEDIAMINE) COBALT(III) COMPLEXES

Submitted by J. SPRINGBØRG† and C. E. SCHÄFFER†
Checked by JOHN M. PRESTON‡ and BODIE DOUGLAS‡

A great majority of the hitherto known dianionobis(ethylene-diamine)cobalt(III) complexes have traditionally been prepared by

*chxn = *trans*-1,2-cyclohexanediamine.

†Chemistry Department I, The H. C. Ørsted Institute, University of Copenhagen, DK-2100 Denmark.
‡University of Pittsburgh, Pittsburgh, Pa. 15213.

syntheses with *trans*-dichlorobis(ethylenediamine)cobalt(III) chloride as the initial starting material. The main reason for using this particular compound is that it has until recently been one of the most available dianionobis(ethylenediamine)cobalt(III) compounds. However, *trans*-dichlorobis(ethylenediamine)cobalt(III) chloride is, by the traditional method, only obtained in a yield of about 50% of a crude product that often contains an impurity of cobalt(II) which is used in excess.[1] Consequently, preparations of dianionobis(ethylenediamine)cobalt(III) complexes that use the *trans*-dichloro salt as the starting material are bound to give low yields based upon original cobalt(II) salt and ethylenediamine.

In the following, a series of preparations of dianionobis(ethylenediamine)cobalt(III) compounds, all starting with (carbonato)bis(ethylenediamine)cobalt(III) chloride are given. (Carbonato)bis(ethylenediamine)cobalt(III) chloride has been prepared with a high yield (80%) by a new method based upon the use of cobalt(II) chloride and the equivalent amount of (2-aminoethyl)carbamic acid. The carbonato compound is easily converted into a number of dianionobis(ethylenediamine)cobalt(III) compounds with high yields. In some of the following procedures it was possible to use the carbon dioxide–ethylenediamine reaction mixture directly.

Besides (carbonato)bis(ethylenediamine)cobalt(III) chloride and the corresponding bromide salt, the following complexes are described: *cis*- and *trans*-dichlorobis(ethylenediamine)cobalt(III) chloride, *cis*-aquachlorobis(ethylenediamine)cobalt(III) sulfate, *cis*-bis(ethylenediamine)dinitrocobalt(III) nitrite, *cis*-aquabis(ethylenediamine)hydroxocobalt(III) dithionate, and *cis*-diaquabis(ethylenediamine)cobalt(III) bromide.

A.   (CARBONATO)BIS(ETHYLENEDIAMINE)COBALT(III) CHLORIDE AND BROMIDE

$$H_2NCH_2CH_2NH_2 + CO_2 \longrightarrow H_2NCH_2CH_2NHCOOH$$
$$2CoCl_2 + 2LiOH + 4H_2N \cdot CH_2 \cdot CH_2 \cdot NH \cdot COOH + H_2O_2 \longrightarrow$$
$$2[Co(en)_2CO_3]Cl + 2CO_2 + 2LiCl + 2H_2O$$

The following procedure is based on the reaction of an aqueous solution of cobalt(II) chloride with the equivalent amount of (2-aminoethyl)carbamic acid, followed by oxidation with hydrogen peroxide and the subsequent formation of bis(ethylenediamine)cobalt(III) ions. The bis(ethylenediamine)cobalt(III) species are converted to the carbonato complex by reaction with lithium hydroxide and carbon dioxide. During the entire preparation a vigorous stream of carbon dioxide is bubbled through the reaction mixture. This procedure appears to be essential in order to minimize the formation of tris(ethylenediamine)cobalt(III) chloride as a by-product. However, the formation of a negligible amount of the tris salt cannot be avoided. The crude salts have a purity suitable for preparative purposes. The pure salts are obtained by recrystallization from aqueous solution.

The optical antipodes have been obtained recently by resolution using the $(+)_D$-(ethylenediamine)bis(oxalato)cobalt(III) anion, and have been isolated as the iodide salt.[2] The exchange reaction with carbonate and the racemization reaction in aqueous solution have been investigated kinetically.[3]

Procedure

A stream of carbon dioxide is bubbled through a mixture of 133 ml. (1.64 moles) of ethylenediamine monohydrate and 133 ml. of water* cooled in ice. The stream of carbon dioxide is maintained during the entire preparation. A solution of 195 g. (0.82 mole) of cobalt(II) chloride hexahydrate in 175 ml. of water at room temperature is added to the cold solution, which is continually stirred. The addition of the cobalt(II) salt causes a violent evolution of carbon dioxide gas, and the solution becomes red-violet. (Sometimes the mixture coagulates and becomes gel-like.) The cooling and the stirring are continued, and the mixture is oxidized by dropwise addition of 200 ml. of 30%

*Alternatively, 112 ml. of 98% ethylenediamine (1.64 moles) and 160 ml. of water may be utilized.

hydrogen peroxide for approximately 45 minutes. (If a gel has been formed, manual stirring is necessary during the addition of approximately the first 50 ml. of hydrogen peroxide, until the mixture again appears to be homogeneous.) During the addition of hydrogen peroxide the temperature increases to about 35°C., and the solution becomes a deeper red. The mixture is heated for about 15 minutes to a temperature range of 70–75°C. and is kept at that temperature for an additional 15 minutes before being filtered and cooled in an ice bath to about 20°C. At this temperature, 34.4 g. (0.82 mole) of finely powdered lithium hydroxide monohydrate is added under a vigorous stream of carbon dioxide, with thorough stirring and no cooling. The temperature rises to about 35°C., and the solution becomes a pure red. The mixture is allowed to remain at room temperature, with constant stirring for half an hour, whereupon fine red crystals of $[Co(en)_2(CO_3)]Cl$ begin to form. Five hundred milliliters of methanol is added, and the mixture is then cooled for 2 hours in an ice bath to effect complete precipitation. It is not advisable to leave the mixture for crystallization overnight, since the yield of the carbonato salt will not increase, but minute amounts of tris(ethylenediamine)cobalt(III) chloride may crystallize out. The stream of carbon dioxide is maintained during the cooling. Cooling without the addition of methanol produces the solution to be used in the preparations of *trans*-dichlorobis(ethylenediamine)-cobalt(III) chloride and *cis*-bis(ethylenediamine)dinitrocobalt(III) nitrite described below, where this solution is referred to as *solution A*. The precipitate is filtered, washed with 200 ml. of 50% v/v ethanol, and dried in air. The yield is 179 g. (80%). The crude product is almost pure. *Anal.* Calcd. for $[Co(en)_2CO_3]Cl$: Co, 21.46; N, 20.40; C, 21.87; H, 5.87; Cl, 12.91. Found: Co, 21.59; N, 20.33; C, 21.61; H, 5.88; Cl, 12.99.

The pure chloride salt is obtained by reprecipitation from water. A 5-g. quantity is dissolved in 30 ml. of water at approximately 90°C. The solution is filtered quickly. To the hot (50–60°C.)

solution is added 30 ml. of methanol with stirring, and the mixture is cooled in an ice bath for one hour. The precipitate is filtered, washed with two 5-ml. portions of 95% ethanol, and allowed to dry in air. The yield is 4.3 g. (86%). The visible absorption spectrum is not changed upon further recrystallizations. *Anal.* Found: Co, 21.45; N, 20.62; C, 21.81; H, 5.90; Cl, 13.05.

The preparation of the bromide salt follows the above synthesis for the chloride salt with minor modifications. With the above quantities, a solution of 268 g. (0.82 mole) of cobalt(II) bromide hexahydrate in 300 ml. of water is substituted for the cobalt(II) chloride solution. With one exception, the above directions are followed exactly. The bromide salt is isolated by cooling in an ice bath without the addition of methanol. The precipitate is filtered, washed with 200 ml. of 50% v/v ethanol and two 200-ml. portions of 95% ethanol, and allowed to dry in air. The yield is 186 g. (71%).* *Anal.* Calcd. for [Co(en)$_2$(CO$_3$)]Br: Co, 18.47; N, 17.56; C, 18.82; H, 5.06; Br, 25.05. Found: Co, 17.65; N, 15.27; C, 17.72; H, 5.10; Br, 26.82.

The pure bromide salt is obtained by recrystallization from water. Four grams is dissolved in 34 ml. of water at 90°C., and the filtered solution is cooled in an ice bath for 2 hours. The precipitate is filtered, washed with 2 ml. of 50% v/v ethanol and two 2-ml. portions of 96% ethanol, and dried in air. The yield is 2.5 g. (63%). The visible absorption spectrum is not changed upon further recrystallization. *Anal.* Found: Co, 18.53; N, 17.54; C, 18.78; H, 5.05; Br, 24.99.

Properties

An aqueous solution of the (carbonato)bis(ethylenediamine)-cobalt(III) ion is rather stable.[3] The visible absorption spectrum of

---

*The checkers report that the same percentage yield is obtained by using quantities scaled to $\frac{1}{5}$ of those specified.

the chloride salt in water showed $(\epsilon,\lambda*)_{max.}$: (132.7, 511.5), (121.1, 359.5); $(\epsilon,\lambda)_{min.}$: (16.1, 427), (41.6, 322). Found for the bromide salt: $(\epsilon,\lambda)_{max.}$: (131.7, 511.5), (120.4, 359.5); $(\epsilon,\lambda)_{min.}$: (15.2, 427), (40.7, 322). The experimental reproducibility of the ϵ values of a definite sample can be characterized by a standard deviation of 0.3%. However, from this point on, when it is stated that a spectrum of a given salt does not change upon further recrystallization, a deviation of 0.8% in the ϵ values (for both maxima and minima) for two consecutive crops has been accepted. The ϵ values noted above for the chloride and the bromide salts are seen to deviate more than 0.8%, especially for the ϵ values at the minima. The bromide salt is believed to be the more nearly pure, since the ratio $\epsilon_{min.}/\epsilon_{max.}$ here is smaller.

B. *trans*-DICHLOROBIS(ETHYLENEDIAMINE)COBALT(III) CHLORIDE

$$[Co(en)_2(CO_3)]Cl + 3HCl + H_2O \longrightarrow$$
$$trans\text{-}[Co(en)_2Cl_2]Cl\cdot2H_2O\cdot HCl + CO_2$$
$$trans\text{-}[Co(en)_2Cl_2]Cl\cdot2H_2O\cdot HCl \xrightarrow{heat}$$
$$trans\text{-}[Co(en)_2Cl_2]Cl + HCl + 2H_2O$$

The traditional method of isolating the *trans*-dichlorobis-(ethylenediamine)cobalt(III) chloride salt by evaporation of the reaction mixture to dryness in the steam bath gives some reduction to cobalt(II). To avoid this, the chloride salt in the following procedure is isolated by saturation of the reaction mixture with hydrogen chloride gas and by precipitation with standing at room temperature. As an alternative method, the reaction mixture from the preparation of (carbonato)bis(ethylene-diamine)cobalt(III) chloride can be used as the starting material.

*λ in nanometers.

Procedure

1. Preparation from (Carbonato)bis(ethylenediamine)cobalt(III) Chloride

A 27.5-g. (0.1-mole) quantity of (carbonato)bis-(ethylenediamine)cobalt(III) chloride is added to 45 ml. of ice-cold 12 M hydrochloric acid in a 100-ml. Erlenmeyer flask with stirring and cooling in an ice bath. The carbonato complex is dissolved with evolution of carbon dioxide gas and formation of a red-violet solution. With continued cooling, the cold (approximately 10°C.) solution is saturated with hydrogen chloride gas, and is then allowed to reach room temperature. The flask is closed and allowed to stand in the dark for 3–4 days. During this time, dark green crystals of *trans*-$[Co(en)_2Cl_2]Cl \cdot 2H_2O \cdot HCl$ separate. The precipitate is filtered from the mother liquor, which is slightly blue from a trace of cobalt(II). Washing three times with 40-ml. portions of 12 M hydrochloric acid and two 50-ml. portions of absolute ethanol and drying at 110°C. yields 27.2 g. (95%) of bright green metamorphs of *trans*-$[Co(en)_2Cl_2]Cl$. *Anal.* Calcd. for *trans*-$[Co(en)_2Cl_2]Cl$: Co, 20.64; N, 19.63; C, 16.83; Cl, 37.26; H, 5.65. Found: Co, 20.67; N, 19.70; C, 16.88; Cl, 37.26; H, 5.50.

2. Preparation from Cobalt(II) Chloride Hexahydrate

To one-fifth of solution A (0.164 mole of cobalt(II) salt, see preparation A) is added 60 ml. of ice-cold 12 M hydrochloric acid. The solution is filtered, cooled to a temperature of 5–10°C., and saturated with hydrogen chloride gas as above. The *trans*-dichloro salt is isolated as before. The yield is 35.1 g.* (75% based upon cobalt(II) chloride). *Anal.* Found: Co, 20.61; N, 19.66; C, 16.89; Cl, 37.25; H, 5.57.

---

*When addition of lithium hydroxide in the preparation of solution A is excluded, the yield of the *trans*-dichloro salt is lowered to 69%.

Properties

The visible absorption spectra of aqueous solutions (extrapolated to $t = 0$) of the samples from the above preparations are identical; $(\epsilon,\lambda)_{max.}$: (37.2, 618); $(\epsilon,\lambda)_{min.}$: (5.1, 525); $(\epsilon,\lambda)_{shoulder}$: (28.2, 455), (37.8, 400).

C. *cis*-DICHLOROBIS(ETHYLENEDIAMINE)COBALT(III) CHLORIDE

$$[Co(en)_2(CO_3)]\,Cl + 2HCl \longrightarrow cis\text{-}[Co(en)_2 Cl_2]\,Cl{\cdot}H_2O + CO_2$$

Procedure

To 27.5 g. (0.1 mole) of crude (carbonato)bis(ethylenediamine)-cobalt(III) chloride is added 200 ml. of 1.00 N hydrochloric acid. The carbonato complex is dissolved with evolution of carbon dioxide gas and formation of a red solution consisting primarily of the corresponding *cis*-diaqua species. The solution is evaporated in the steam bath until an almost dry paste has been formed. The purple residue is filtered and washed with three 20-ml. portions of ice-cold water. Drying in air yields 19.5 g. of purple crystals of *cis*-dichlorobis(ethylenediamine)cobalt(III) chloride. The mother liquor and the washings are again evaporated almost to dryness to yield a second crop of crystals, 5.9 g. The total yield is 25.4 g. (84% based on (carbonato)bis(ethylenediamine)cobalt(III) chloride). The analysis and the visible absorption spectrum of the two fractions are identical. *Anal.* Calcd. for $[Co(en)_2 Cl_2]\,Cl{\cdot}H_2O$: Co, 19.42; N, 18.46; C, 15.82; Cl, 35.05; H, 5.98. Found: Co, 19.50; N, 18.57; C, 15.77; Cl, 35.15; H, 6.01.

Properties

The spectrum was measured in 12 M hydrochloric acid to prevent hydrolysis. However, even in this medium an extrapolation of the spectrum back to time of dissolution (a correction of about 1%) is required because of the isomerization reaction, $(\epsilon, \lambda)_{max}$: (98.9, 534.5); $(\epsilon, \lambda)_{min}$: (16.6, 451); $(\epsilon, \lambda)_{shoulder}$: (91.4, 388). For other properties, see reference 1.

D. *cis*-AQUACHLOROBIS(ETHYLENEDIAMINE)COBALT(III) SULFATE

$$[Co(en)_2(CO_3)]Cl + H_2SO_4 + 2H_2O \longrightarrow$$
$$cis\text{-}[Co(en)_2(H_2O)Cl]SO_4 \cdot 2H_2O + CO_2$$

The bromide or chloride salt can be prepared easily from the sulfate salt.[4b,5] Werner has resolved the bromide salt into its optical antipodes through the (3-bromo-2-oxo-8-bornanesulfonic acid).[6] The bromide salt is converted by heat into a mixture of the corresponding *cis*- and *trans*-bromochloro complexes.[5]

Procedure

A 41.2-g. (0.15-mole) sample of crude (carbonato)bis-(ethylenediamine)cobalt(III) chloride is added, with caution, portionwise to 40 ml. of 4 M sulfuric acid in a 500-ml. conical flask at room temperature. The carbonato complex dissolves with evolution of carbon dioxide gas and formation of a red solution of the corresponding *cis*-diaqua species. The flask is fitted with a condenser, and the mixture is then heated in the water bath at 85°C. for $\frac{1}{2}$ hour. The solution turns from red to red-violet. The reaction mixture is transferred from the water bath to an ice bath and cooled to room temperature. Without cooling, 20 ml. of 96% ethanol is added to the stirred solution. The Erlenmeyer flask is closed, and after 1–2 hours the precipitation of red-violet crystals of *cis*-aquachlorobis(ethylenediamine)cobalt(III) sulfate commences. To complete the precipitation, the reaction mixture is allowed to remain at room temperature in darkness with continued stirring for 3–4 days. The precipitate is filtered, washed thoroughly with three 45-ml. portions of 33% v/v ethanol and two 75-ml. portions of 96% ethanol. The bis(ethylenediamine)cobalt(III) in the mother liquor and the washings may be recovered as *trans*-dichlorobis-(ethylenediamine)cobalt(III) chloride. Before doing this, it is advisable to remove the ethanol at low temperature in order to minimize the reduction to cobalt(II). Drying in air yields 39 g. (71%) of the almost pure sulfate salt. *Anal.* Calcd. for $[Co(en)_2(H_2O)Cl]SO_4 \cdot 2H_2O$: Co, 16.16; C, 13.17; N, 15.36; Cl, 9.72; H, 6.08. Found: Co, 16.19; C, 13.10; N, 15.41; Cl, 9.51; H, 6.07.

Properties

The sulfate is sparingly soluble in water. In aqueous solution, it hydrolyzes much more slowly than does the corresponding *cis*-dichloro complex with respect to substitution of its first chloride by water. The visible spectrum in water showed $(\epsilon,\lambda^*)_{max.}$: (87.9, 516); (72, 375). This spectrum is in fair agreement with that reported in the literature[7] for the pure chloride salt, $(\epsilon,\lambda^*)_{max.}$: (85.5, 516).

E.   *cis*-BIS(ETHYLENEDIAMINE)DINITROCOBALT(III) NITRITE

$$[Co(en)_2(CO_3)]^+ + 2H^+ + H_2O \longrightarrow$$
$$cis\text{-}[Co(en)_2(H_2O)_2]^{3+} + CO_2$$
$$cis\text{-}[Co(en)_2(H_2O)_2]^{3+} + 3NO_2^- \longrightarrow$$
$$cis\text{-}[Co(en)_2(NO_2)_2]NO_2 + 2H_2O$$

Procedure

A 137.3-g. (0.5-mole) sample of crude (carbonato)bis(ethylenediamine)cobalt(III) chloride is added with stirring to 275 ml. of 4 *M* hydrochloric acid cooled to 0–5°C. in an ice bath. The

*λ in nanometers.

carbonato complex dissolves with evolution of carbon dioxide gas and formation of a red solution of the corresponding *cis*-diaqua complex. The cooling is maintained for about 20 minutes, after which period the solution is filtered. A hot solution (70–72°C.) composed of 550 g. (8.0 moles) of sodium nitrite dissolved in 475 ml. of water is added to the cold red solution (5–10°C.) of the *cis*-diaqua complex. The addition is made carefully over a period of 1–2 minutes without further cooling, but with adequate stirring. Because of vigorous frothing a large beaker (4–6 l.) is used for the mixing. The solution turns from red to orange-brown, and at the beginning vigorously produces nitrogen oxides. During the mixing, yellow crystals of *cis*-bis(ethylenediamine)dinitrocobalt(III) nitrite begin to precipitate. The temperature is 50–55°C. immediately after the mixing. The reaction mixture is kept at that temperature for 15 minutes and then cooled in an ice bath for 3 hours to complete the precipitation. The precipitate is filtered, washed with one 100-ml. portion of ice-cold water and two 100-ml. portions of 96% ethanol, and allowed to dry in air. The yield is 136 g. (86%) of *cis*-bis(ethylenediamine)dinitrocobalt(III) nitrite. *Anal.* Calcd. for $[Co(en)_2(NO_2)_2]NO_2$: Co, 18.58; C, 15.15; H, 5.09; N, 30.92. Found: Co, 18.80; C, 15.13; H, 5.09; N, 30.95.

Preparation from Cobalt(II) Chloride

One hundred fifty milliliters of 12 M hydrochloric acid is added to a solution containing 195 g. (0.82 mole) of cobalt(II) chloride hexahydrate in 175 ml. of water. The cobalt(II) chloride solution is cooled in an ice bath and stirred during addition of the acid. This solution is treated with a solution of 900 g. (13.1 moles) of sodium nitrite in 775 ml. of water, and *cis*-bis(ethylenediamine)dinitrocobalt(III) nitrite is isolated as described above. The yield is 178 g. of the impure nitrite salt (68.5% based on cobalt(II) chloride). *Anal.* Found: Co, 18.98; C, 14.85; H, 5.08; N, 29.51.

The cis configuration has been confirmed by conversion of the crude nitrite salts into the bromide salt and resolution of the optical antipodes that are formed, according to the method of Dwyer and Garvan's.[8]

Properties are reported in previous volumes in this series.[9,10]

F. *cis*-AQUABIS(ETHYLENEDIAMINE)HYDROXO-COBALT(III) DITHIONATE

$$[Co(en)_2(CO_3)]^+ + 2H^+ + H_2O \longrightarrow$$
$$cis\text{-}[Co(en)_2(H_2O)_2]^{3+} + CO_2$$
$$cis\text{-}[Co(en)_2(H_2O)_2]^{3+} + OH^- + [S_2O_6]^{2-} \longrightarrow$$
$$cis\text{-}[Co(en)_2(OH)(H_2O)][S_2O_6] + H_2O$$

Procedure

A 16.5-g. (0.06-mole) sample of crude (carbonato)bis-(ethylenediamine)cobalt(III) chloride is dissolved in 40.0 ml. of 2 M sulfuric acid at room temperature. The carbonata complex dissolves, producing carbon dioxide gas and a red solution of the corresponding *cis*-diaqua complex. To remove all the carbon dioxide, a stream of nitrogen gas is bubbled through the solution, which is cooled in an ice bath for one hour. This procedure is necessary to avoid partial formation of the carbonato complex by the subsequent addition of sodium hydroxide. A solution of 14.6 g. (0.06 mole) of sodium dithionate dihydrate in 175 ml. of water (20°C.) is added, and the filtered solution is cooled to 5–10°C. Now 50 ml. of ice-cold, 2 N sodium hydroxide is added to the stirred solution with continued cooling. Microscopic pink crystals of *cis*-aquabis(ethylenediamine)-hydroxocobalt(III) dithionate are immediately precipitated. After the suspension has cooled for another 5 minutes, the precipitate is filtered and washed with four 25-ml. portions of water and 96% ethanol. Drying in air yields 15.9 g. (71%) of the almost pure dithionate salt. *Anal.* Calcd. for $[Co(en_2)(OH)(H_2O)]S_2O_6$: Co, 15.75; C, 12.84; N, 14.97; H, 5.12. Found: Co, 15.96; C, 12.74; N, 15.09; H, 5.12.

The pure sample is obtained by reprecipitation. Ten grams of the crude product is dissolved in 60.0 ml. of ice-cold 0.5 N hydrochloric acid with cooling in an ice bath. Then 15 ml. of ice-cold 2 N sodium hydroxide is added to the filtered solution with stirring and cooling, and *cis*-aquabis(ethylenediamine)-hydroxocobalt(III) dithionate is isolated as above. Yield is 8.4 g. (84%). The visible absorption spectrum in acidic or basic solution of the sample reprecipitated twice in this manner does not change after the second precipitation. *Anal.* Found: Co, 15.74; C, 12.72; N, 14.87; H, 5.10.

Properties

The dithionate salt is almost insoluble in water. It dissolves in acid or base with formation of the corresponding *cis*-diaqua and *cis*-dihydroxo species, respectively.[11] The absorption spectrum in 0.12 M perchloric acid showed $(\epsilon,\lambda)_{max}$: (78.5, 492), (61.5, 358); $(\epsilon,\lambda)_{min}$: (13.6, 414), (10.2, 305). The absorption spectrum in 0.1 N sodium hydroxide showed $(\epsilon,\lambda)_{max}$: (94.4, 517), (102.3, 371); $(\epsilon,\lambda)_{min}$: (15.8, 433), (19.0, 319.5).

In 1 M sodium nitrate, the acidity constants for *cis*-diaqua and *cis*-aquahydroxo ions are $pK_1 = 6.06$ and $pK_2 = 8.19$.[11]

G.   *cis*-DIAQUABIS(ETHYLENEDIAMINE)COBALT(III) BROMIDE

$$cis\text{-}[Co(en)_2\,(OH)(H_2O)]\,[S_2O_6\,] + 3HBr + H_2O \longrightarrow$$
$$cis\text{-}[Co(en)_2\,(H_2O)_2\,]\,Br_3\cdot H_2O + H_2S_2O_6$$

Procedure

Twenty grams (0.0535 mole) of crude *cis*-aquabis(ethylene-diamine)hydroxocobalt(III) dithionate (preparation F) is dissolved in 60 ml. of ice-cold (0–5°C.) 1 M hydrobromic acid. To the filtered solution is added, with stirring and cooling in an ice bath, 120 ml. of ice-cold concentrated hydrobromic acid* (sp. gr., 1.73). After some minutes red crystals of *cis*-diaquabis(ethylenediamine)cobalt(III) bromide separate. The cooling is continued for half an hour to complete the precipitation. The sample is filtered, washed thoroughly with 96% ethanol, and dried in an evacuated desiccator over 5 *M* sulfuric acid at 5°C. This procedure yields 21.0 g. (80%) of red crystals of the almost pure bromide salt.* *Anal.* Calcd. for $[Co(en)_2(H_2O)_2]Br_3 \cdot H_2O$: Co, 12.46; N, 11.85; C, 10.16; Br, 50.70. Found: Co, 12.39; N, 11.73; C, 9.88; H, 4.87; Br, 49.97.

Properties

To prevent substitution by bromide, the sample is kept in a refrigerator at −15°C. At this temperature it is stable for some months. The above crude product of the bromide salt is contaminated with a negligible amount of *cis*-aquabromobis-(ethylenediamine)cobalt(III) bromide, which is not removed by reprecipitation. However, the bromide salt has a purity suitable for further synthetic work, e.g., preparation of *cis*-aquabis(ethylenediamine)hydroxocobalt(III) bromide.[12]

The visible absorption spectrum in acidic solution differs slightly from the spectrum of the pure *cis*-diaqua complex.† Found in 0.12 *M* perchloric acid: $(\epsilon,\lambda)_{max.}$, (76.3, 492), (60.8, 358); $(\epsilon,\lambda)_{min.}$, (13.1, 414), (22.5, 313).

References

1. J. C. Bailar, Jr., *Inorganic Synthesis,* 2, 222 (1946).
2. F. P. Dwyer, A. M. Sargeson, and I. K. Reid, *J. Am. Chem. Soc.,* 85, 1215 (1963).
3. J. S. Holden and G. M. Harris, *ibid.,* 77, 1934 (1955).

*The checkers report a lower yield of product (60%) because they used 48% HBr (sp. gr., 1.48).
†See preparation F.

4. A. Werner, (a) *Ann. Chem.*, **386**, 1 (1912); (b) *ibid.*, p. 123.
5. J. W. Vaughn and R. D. Lindholm, *Inorganic Syntheses*, **9**, 163 (1967).
6. A. Werner, *Helv. Chim. Acta*, **4**, 113 (1921).
7. P. Benson and A. Haim, *J. Am. Chem. Soc.*, **87**, 3826 (1965).
8. F. P. Dwyer and F. L. Garvan, *Inorganic Syntheses*, **6**, 195 (1960).
9. E. P. Harbulak and M. J. Albinak, *ibid.*, **8**, 196 (1966).
10. H. F. Holtzclaw, Jr., D. P. Sheetz, and B. D. McCarty, *ibid.*, **4**, 176 (1953).
11. J. Bjerrum and S. E. Rasmussen, *Acta Chem. Scand.*, **6**, 1265 (1952).
12. A. Werner, *Chem. Ber.*, **40**, 272 (1907).

15. BIS(ETHYLENEDIAMINE)SULFITO COMPLEXES OF COBALT(III)

Submitted by ROBERT D. HARGENS,* WOONZA MIN,* and ROBERT C. HENNEY*
Checked by T. M. BROWN† and A. GALLIART†

Disodium *cis*-bis(ethylenediamine)disulfitocobaltate(III) nitrate and also the perchlorate have not been reported before and are recommended as sources of the *cis*-bis(ethylenediamine)disulfitocobaltate(III) ion, which has been prepared by Baldwin[1] by proceeding through a difficult series of complex salts. The *cis*-azidobis(ethylenediamine)sulfitocobalt(III) is also newly reported; it can be converted into a relatively pure sodium *trans*-bis(ethylenediamine)disulfitocobaltate(III), which has previously been made from dichlorobis(ethylenediamine)cobalt(III) chloride[1] and tetraammine(carbonato)cobalt(III) chloride.[2]

A. DISODIUM *cis*-BIS(ETHYLENEDIAMINE)DISULFITOCOBALT(III) NITRATE AND PERCHLORATE

$$2[Co(H_2O)_6](NO_3)_2 + 4Na_2SO_3 + 4en + O_2 + 2HNO_3 \longrightarrow$$
$$2Na_2[Co(en)_2(SO_3)_2]NO_3 \cdot 3H_2O + H_2O_2 + 4NaNO_3 + 6H_2O$$
$$Na_2[Co(en)_2(SO_3)_2]NO_3 \cdot 3H_2O + NaClO_4 \cdot H_2O \longrightarrow$$
$$Na_2[Co(en)_2(SO_3)_2]ClO_4 \cdot 3H_2O + NaNO_3 + H_2O$$

*Mankato State College, Mankato, Minn. 56001.
†Arizona State University, Tempe, Ariz. 85281.

Procedure

A mixture of 25 g. (0.20 mole) of sodium sulfite and 100 ml. of water is added to 29 g. (0.10 mole) of cobalt(II) nitrate 6-hydrate dissolved in 25 ml. of water, which gives a pink, mushy precipitate of cobalt(II) sulfite. Then 6 ml. of concentrated nitric acid in 10 ml. of water is added dropwise to 12 g. (0.20 mole) of 98% ethylenediamine and mixed with the cobalt sulfite, which immediately dissolves to form a brown solution. (■ *Caution. Spattering occurs during the addition of the nitric acid.*) Air is bubbled through the solution with a wash-bottle attachment for one hour, and the solution is filtered to remove any residue of cobalt oxide.

The filtrate may be treated with 15 g. (0.11 mole) of sodium perchlorate 1-hydrate to obtain light-brown crystals of disodium *cis*-bis(ethylenediamine)disulfitocobaltate(III) perchlorate 3-hydrate, or it may be concentrated under reduced pressure in a flash evaporator until small crystals or a yellow powder of disodium *cis*-bis(ethylenediamine)disulfitocobaltate(III) nitrate 3-hydrate appear. In either case the solution is cooled in an ice bath for one hour before filtering; the crystals are washed with alcohol and ether and dried at 125°C. At this temperature each salt gives up three water molecules and darkens in color. Since the crude nitrate is difficult to recrystallize, the nearly pure perchlorate salt is preferable for most purposes. Yield of the anhydrous nitrate salt is 4–7 g. (9–15%). *Anal.* Calcd. for $Na_2[Co(en)_2(SO_3)_2]NO_3$: Co, 13.2; Na, 10.3; NO_3^-, 13.9. Found: Co, 12.5; Na, 10.1; NO_3^-, 14.8. Yield of the anhydrous perchlorate salt is 25–30 g. (52–62%). *Anal.* Calcd. for $Na_2[Co(en)_2(SO_3)_2]ClO_4$: Co, 12.2; N, 11.6; Na, 9.5. Found: Co, 12.1; N, 11.6; Na, 9.5.

B.   *cis*-AZIDOBIS(ETHYLENEDIAMINE)SULFITOCOBALT(III)

$cis\text{-}[Co(en)_2(N_3)_2]NO_3 + Na_2SO_3 \longrightarrow$
$$cis\text{-}[Co(en)_2N_3(SO_3)] + NaN_3 + NaNO_3$$

Procedure

A mixture of 20 g. (0.062 mole) of *cis*-diazidobis(ethylenediamine)cobalt(III) nitrate[3,]* and 40 ml. of warm water is placed in a steam bath. Sodium sulfite 7.8 g. (0.062 mole) is stirred slowly into the mixture, and stirring is continued for 2 minutes. The hot solution is filtered quickly to remove cobalt oxide. Red crystals of *cis*-azidobis(ethylenediamine)sulfitocobalt(III) start to precipitate immediately and, after cooling in an ice bath for one hour, they are filtered. After recrystallization from boiling water (20 ml./g. of crude product), the yield is 4–5 g. (22–27%). If it is desired to prepare sodium *trans*-bis(ethylenediamine)disulfitocobaltate(III), one may triple the amounts of reagents to obtain a sufficient yield. *Anal.* Calcd. for $[Co(en)_2 N_3 (SO_3)]$: Co, 19.6. Found: Co, 19.5.

C. SODIUM *trans*-BIS(ETHYLENEDIAMINE)DISULFITO-COBALTATE(III)

$$cis\text{-}[Co(en)_2 N_3 (SO_3)] + Na_2 SO_3 + 3H_2 O \longrightarrow$$
$$trans\text{-}Na[Co(en)_2 (SO_3)_2] \cdot 3H_2 O + NaN_3$$

Procedure

A sample of 11.6 g. (0.039 mole) of *cis*-azidobis(ethylenediamine)sulfitocobalt(III) is dissolved in 15 ml. of water. The solution is heated to boiling, and 5 g. (0.040 mole) of sodium sulfite is slowly stirred in. After being boiled for 3 minutes, the solution is cooled in an ice bath. Bright yellow crystals of sodium

---

*The checkers suggest that in following the directions of Staples and Tobe[3] for the preparation of this salt, the concentrated HNO_3 be diluted with an equal volume of water before adding it to the ethylenediamine.

trans-bis(ethylenediamine)disulfitocobaltate(III)   3-hydrate   form
immediately and after a few minutes are filtered off, washed with
ethanol and ether, and dried at 120°C., at which temperature the
three water molecules are driven off. The yield is 7–8 g. (50–57%).
Anal. Calcd. for $Na[Co(en)_2(SO_3)_2]$: Co, 16.3; N, 15.5; Na, 6.4.
Found: Co, 16.1; N, 15.6; Na, 6.7.

Properties

Polarographic analysis of the *cis*-bis(ethylenediamine)disulfito-
cobaltate(III) ion indicates a sulfite/cobalt ratio of 2 and identifies
the complex ion as that in the compound described by Baldwin as
sodium *cis*-bis(ethylenediamine)disulfitocobaltate(III).   Baldwin's
tentative assignment of the cis configuration is supported by
lability studies by Stranks and Yandell.[4] The ion has an
absorption maximum at 442 nm. (440 reported by Baldwin). The
x-ray diffraction pattern of the nitrate salt does not disclose any
evidence of sodium nitrate crystals. The nitrate can be converted
to the perchlorate by dissolving it in hot water, adding sodium
perchlorate, and cooling. The equivalent weight of the anhydrous
perchlorate salt determined by cation exchange is 238 (calcd.,
242). The anhydrous salts rapidly absorb three water molecules
when exposed to the atmosphere.

The *cis*-azidobis(ethylenediamine)sulfitocobalt(III) exhibits an
absorption maximum at 502 nm. The infrared spectrum shows
expected bands for the sulfito and azido groups. Polarographic
analysis indicates a cobalt/sulfite ratio of 1. The assignment of the
cis configuration is tentative, based on the generally complex
infrared spectrum.[5]

The sodium *trans*-bis(ethylenediamine)disulfitocobaltate(III) is
identical in all respects with the product prepared by Baldwin's
method, although the absorption peak for the product obtained
by either method is observed at 439 nm., compared with 431
reported by Baldwin. The complex ion undergoes rapid acid or
base hydrolysis to form the aquabis(ethylenediamine)sulfito-

cobalt(III) ion. If recrystallization is indicated, losses caused by hydrolysis can be minimized by recrystallizing from 0.5 M sodium sulfite solution.

References

1. M. E. Baldwin, *J. Chem. Soc.*, **1961**, 3123.
2. R. Klement, *Z. Anorg. Allgem. Chem.*, **150**, 117 (1926).
3. P. J. Staples and M. L. Tobe, *J. Chem. Soc.*, **1960**, 4812.
4. D. R. Stranks and J. K. Yandell, *Inorg. Chem.*, **9**, 751 (1970).
5. M. L. Morris and D. H. Busch, *J. Amer. Chem. Soc.*, **82**, 1521 (1960).

16. NITROSYLIRON, –COBALT, AND –NICKEL IODIDES

Submitted by B. HAYMORE* and R. D. FELTHAM†
Checked by B. E. MORRIS‡ and R. A. CLEMENT‡

The compounds $[Fe(NO)_2 I]_2$, $[Co(NO)_2 I]_x$, and $[Ni(NO)I]_x$ have been prepared before with gas-solid reactions at elevated temperatures.[1] However, the syntheses were complicated and the yields were relatively low, expecially for the cobalt and nickel compounds. In addition, the syntheses of these compounds required equipment not available in every laboratory. The syntheses described herein require only readily available glassware and chemicals. If desired, these compounds can be prepared in macro quantities by the methods described below.

*Chemistry Department, Northwestern University, Evanston, Ill. 63301.
†Chemistry Department, University of Arizona, Tucson, Ariz. 85721.
‡E. I. du Pont de Nemours & Company, Wilmington, Del. 19898.

Procedure

A. IRON DINITROSYL IODIDE

(Iododinitrosyliron)

$$Fe + I_2 \xrightarrow{(CH_3)_2CO} FeI_2$$

$$FeI_2 + Fe + 4NO \xrightarrow{(CH_3)_2CO} [Fe(NO)_2I]_2$$

The entire synthesis of the iron dinitrosyl iodide can be carried out in a 1000-ml., three-necked, round-bottomed flask equipped with a gas inlet, a 500-ml., pressure-equalizing addition funnel, and a mechanical stirrer as shown in Fig. 8. A magnetic stirrer must not be used during the preparation of FeI_2 and $[Fe(NO)_2I]_2$, for the large excess of iron powder will prevent efficient stirring.* Because of the toxic nature of nitric oxide, the entire reaction should be carried out in an efficient hood. For best results, all joints should be lubricated with high-vacuum silicone grease. Regulated cylinders of nitrogen and nitric oxide† should be

*The checkers find that a reflux condenser is desirable.
†Matheson Gas Products, East Rutherford, N.J. 07073.

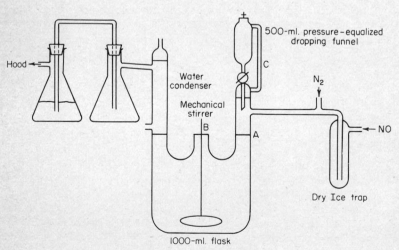

Fig. 8 Apparatus for the preparation of iron dinitrosyl iodide.

attached as indicated, by polyvinyl tubing. Gases exit through the water condenser and the bubble tube, which is attached to the condenser with polyvinyl tubing, or through the unstoppered dropping funnel, as required. Care must be taken that air does not reenter the flask through this bubble trap during subsequent operations.

A strong nitrogen flow is set, the dropping funnel is removed, and to the flask are added anhydrous-grade acetone (250 ml.), deoxygenated by bubbling dry nitrogen through it, and 33.5 g. (0.3 mole) of reagent-grade iron powder. The dropping funnel is replaced, and the system is purged with nitrogen, the gas exiting through the unstoppered dropping funnel and the bubble tube. A solution containing 38.1 g. (0.1 mole) of reagent-grade iodine in 300 ml. of anhydrous, deoxygenated, peroxide-free ether is added to the addition funnel. The addition funnel is then stoppered and the nitrogen flow reduced.

The iodine solution is added dropwise at such a rate that the entire quantity of iodine is added over a period of 45–74 minutes. If the iodine is added too rapidly, iodination of acetone will take place. ■ *Caution. Iodoacetone species are extremely powerful lacrimators.* During the addition of the iodine solution, a very slow flow of nitrogen should be maintained (1–5 bubbles per minute). The resultant product should consist of a dark brown solution and unreacted iron (150-mole % excess). This excess iron powder is necessary for the over-all reaction, for reducing the iodination of acetone, and for improving the yield of product.

After the final addition of the iodine solution, the nitrogen gas is displaced by nitric oxide. The reaction of the nitric oxide is immediate. The flow of this gas should be maintained through the bubble trap so that there is always a positive pressure of nitric oxide inside the flask. A very large flow of nitric oxide can be maintained into the flask at high stirring rates, but only a few bubbles of nitric oxide need to be passed through the bubble trap (*ca.* one bubble per second). However, when the stirring rate is decreased, the flow of nitric oxide must be decreased simultaneously or else large amounts of nitric oxide will be wasted. Vigorous stirring should be maintained throughout the addition of the nitric oxide, because the rate of nitric oxide consumption is dependent upon the rate of mixing between the gas and liquid. Depending upon the stirring rate, the reaction with nitric oxide will take between 60 and 120 minutes. The end of the reaction is marked by the cessation of absorption of nitric oxide, determined by use of the bubble trap. When the reaction is complete, the nitric oxide is displaced by nitrogen. Care must be taken in handling this solution, since in solution the iron nitrosyls are very sensitive to water, oxygen, and especially nitrogen dioxide (from air oxidation of nitric oxide).

Under a flow of nitrogen, a magnetic stirring bar is added to the flask, while the dropping funnel and mechanical stirrer are replaced with glass stoppers. The Dry Ice trap is replaced with a 1000-ml. filtering flask immersed in liquid nitrogen, and the nitric oxide cylinder is replaced with a mechanical vacuum pump. The nitrogen inlet and exit to the bubble trap are clamped off, and the ether and acetone are removed carefully by vacuum distillation into the filter flask. These ether and acetone distillates should be handled with care since they will contain small amounts of a powerful lacrimator.

After the black solid is completely dry, determined by the lack of further condensate in the filter flask and by the formation of a black sublimate in the flask, the flask is filled with nitrogen. The filter flask is replaced with a small liquid-nitrogen trap. The center stopper is replaced with a Pyrex glass cold finger, constructed by making a test-tube end on the bottom of a ground joint with a reduced-diameter lower tube (Fig. 9). The flask is immersed in a 3-l. beaker containing enough mineral oil to come to just below the necks of the 1-l. flask. The beaker and flask are placed upon a magnetic-stirring hot plate where the oil bath is stirred magnetically.

The remaining traces of acetone and other volatile products are removed by vacuum distillation into the liquid-nitrogen trap by

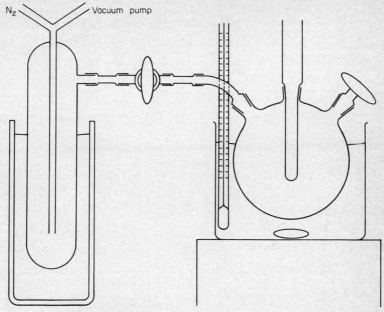

Fig. 9 Sublimation apparatus.

raising the temperature of the mineral oil to 70°C. It is absolutely
essential to handle this higher-boiling fraction in an efficient hood,
since it contains most of the lacrimator produced. If all the
volatile components are not removed from the flask before the
cold finger is charged with carbon dioxide, these liquids will reflux
on the cold finger and prevent sublimation of the product. When
this happens, remove the solid carbon dioxide from the cold
finger, and continue the distillation into the liquid-nitrogen trap
until the solid is completely free of these liquids. The heating rate
of the flask should be approximately 2°C./minute. This slow
heating rate is necessary to ensure uniform heating of the crude
product. *Explosive decomposition* can take place (and has, in one
case) when the heating rate is excessive. The temperature of the
mineral oil is increased until a dark brown or black solid appears in
the liquid-nitrogen trap (*ca.* 90°C.). Solid carbon dioxide is then
added to the cold finger, and a small quantity of acetone is added

to improve contact between the Dry Ice and the walls of the cold finger. The bath temperature is slowly increased until a large amount of sublimate begins to appear on the cold finger (*ca.* 110°C.). Because of the large quantities involved, two batches of sublimate must be collected. The product can be scraped from the cold finger into a bottle in air, providing the bottle is flushed before and after filling it with dry nitrogen. Although solutions of the compound are very sensitive to air, the sublimed solid may be handled for short periods of time in dry air. Alternatively, the flask can be cooled, placed in a dry-bag or dry-box, and the solid transferred from the cold finger to a convenient storage bottle.

The sublimation gives dense, black, lustrous crystals with metallic reflecting surfaces. The yield is 59.8 g. (or 82%) based on iodine. The product was identified by its mass spectrum,[2] infrared spectrum, and elemental analysis. *Anal.* Calcd. for $[Fe(NO)_2 I]_2$: Fe, 23.0; I, 52.3; mol. wt. (monoisotopic), 486. Found: Fe, 23.4; I, 51.8; mol. wt., 486. Infrared spectrum, observed: ν_{NO}, 1810 and 1770 cm.$^{-1}$; in literature: ν_{NO}, 1818 and 1771 cm.$^{-1}$.

B. COBALT DINITROSYL IODIDE
(Iododinitrosylcobalt)

$$Co + I_2 \xrightarrow{(CH_3)_2 CO} CoI_2$$
$$Co + CoI_2 + 4NO \xrightarrow{(CH_3)_2 CO} 2[Co(NO)_2 I] *$$

The cobalt analog of the iron dinitrosyl iodide is prepared according to the above procedure with only minor modifications, which are outlined below. Anhydrous deoxygenated reagent-grade acetone and 35.4 g. of cobalt powder† are added to the 1000-ml. flask and allowed to react with the ether solution of iodine (38.1 g. (0.15 mole) of iodine in 300 ml. of ether) over a period of 30 minutes. Then nitric oxide is admitted to the flask. The reaction

*Equation as represented is balanced for the monomer.
†Fischer Scientific Company, Burlingame, Calif. 94010.

of nitric oxide with cobalt(II) iodide is complete after 30 minutes. After removal of all the volatile liquids by vacuum distillation, the temperature of the flask is raised slowly to $100°C$. Almost no sublimation of the cobalt complex takes place at this temperature in contrast with the iron dinitrosyl iodide, which will entirely sublime under these conditions. The majority of the cobalt dinitrosyl iodide must be sublimed by using a bath temperature between 130 and $140°C$. The cobalt compound sublimes onto the cold finger, cooled to $-78°C$., as lustrous black crystals similar in appearance and crystal habit to those of the iron complex. However, if the cold finger is maintained at $0°C$. with ice, then the cobalt complex sublimes in the form of fine brown needles. This brown phase also appears on the surface of the black sublimate at $-78°C$. when the black sublimate is thick enough to serve as insulation for the cold finger. When the black, low-temperature modification is allowed to warm to room temperature on the probe, it converts to the brown form. The conversion can be easily followed visually. The cobalt complex can also be sublimed by using boiling methanol in the cold finger. Under these conditions, a third high-temperature black form results. On cooling to room temperature, this high-temperature black form also reverts to the brown form. These phase changes are reversible, and the same phenomena can be observed after several sublimations. The crystal structure of the brown form has been determined.[3] This brown modification consists of infinite chains of Co—I—Co—I—, etc. In solution and in the gas phase this cobalt nitrosyl iodide is dimeric, as is the iron complex. It would appear that the black low-temperature modification probably consists of the dimeric species which is deposited directly from the gas phase, whereas the structure of the high-temperature modification is unknown. The yield of $Co(NO)_2I$ is 33.8 g. (or 46%) based on iodine. *Anal.* Calcd. for $[Co(NO)_2I]_2$: Co, 23.97; mol. wt. (monoisotopic), 492. Found: Co, 23.4; mol. wt. (mass spectrometry), 492.*

---

*Molecular weight of dimer form.

C. NICKEL NITROSYL IODIDE

   (Iodonitrosylnickel)

$$Ni + I_2 \xrightarrow{(CH_3)_2 CO} NiI_2$$

$$Ni + NiI_2 + 2NO \xrightarrow{(CH_3)_2 CO} 2[Ni(NO)I]^*$$

The nickel complex, [Ni(NO)I], can also be prepared by the procedures outlined above. The compound is formed readily in reasonable yields, but it is significantly less stable than the iron and cobalt compounds. Moreover, although some of it can be sublimed, it decomposes slowly even below its sublimation temperature. The gram quantities used are the same as for cobalt, and no modification of the procedure is necessary until the sublimation step. No sublimation of the nickel complex takes place until the bath temperature reaches 155–165°C. At this temperature a small amount (2.5 g.) of product sublimes onto the cold probe. *Anal.* Calcd. for Ni(NO)I: Ni, 27.22; I, 58.86. Found: Ni, 26.8; I, 56.5.

Owing to the low volatility and decomposition of the product, the following procedure is preferred. After removal of all the ether and acetone by vacuum distillation, the flask is equipped with a reflux condenser and gas outlet under a strong nitrogen flow. Next, 300 ml. of dry, deoxygenated, *peroxide-free* tetrahydrofuran† is added to the flask, and the mixture is refluxed for 2 hours. After cooling to room temperature, the solution is decanted into a Schlenk tube[4] filled with dry nitrogen. This compound cannot be exposed at any time to oxygen or water. The solid must be transferred by the techniques described above or in a dry-bag or dry-box. The solution is cooled to −78°C. and yields a large amount of a green powder. Green crystals are obtained upon slow

---

*Equation balanced for the monomer.
†Details for the purification of tetrahydrofuran are given in *Inorganic Syntheses*, **12**, 317 (1970).

crystallization at $-78°C$. (24 hours). (The checkers found that the solution supercools and that crystallization is slow.) The solution is quickly filtered through a glass frit of coarse porosity under nitrogen. (The checkers removed the solvent under N_2 flow by use of a syringe.) The green solid is a tetrahydrofuran complex which is stable in the absence of air and water and in the presence of the equilibrium vapor pressure of its tetrahydrofuran. However, all the tetrahydrofuran can be removed from this green solid by heating the solid in a vacuum to $50°C$., as evidenced by the lack of carbon and hydrogen in the sample ($<2\%$). Further material can be obtained by a second extraction, which gives a total yield of Ni(NO)I of 31.8 g. (or 49%) based on iodine.

Properties

The iron compound readily sublimes and yields well-formed, black lustrous crystals. The cobalt complex will also readily sublime, but dependent upon the temperature at which the crystals are formed, they can be either black or brown in color. The crystal structures of both the cobalt and iron complexes have been determined.[3] The nickel complex sublimes only in small amounts with difficulty. All three complexes are unstable to air and water, and the nickel complex readily undergoes thermal decomposition above $100°C$. All three compounds will also readily form complexes with a variety of donor ligands such as tertiary arsines or phosphines. The nickel compound usually forms 2:1 adducts such as $[(C_6H_5)_3P]_2Ni(NO)I$, while the iron and cobalt complexes often undergo disproportionation.[5]

References

1. W. Hieber and R. Nast, *Z. Anorg. Allgem. Chem.*, **224**, 23 (1940).
2. J. Lewis and B. F. G. Johnson, *Acc. Chem. Res.*, **1**, 245 (1968).
3. L. Dahl, E. R. deGil, and R. D. Feltham, *J. Am. Chem. Soc.*, **91**, 1653 (1969).
4. J. J. Eisch and R. B. King, "Organometallic Syntheses," Vol. 1, Academic Press, Inc., New York, 1965.
5. See, for example, B. F. G. Johnson and J. A. McCleverty, *Prog. Inorg. Chem.*, **7**, 277 (1966).

17. POTASSIUM TRICHLORO(ETHYLENE)PLATINATE(II)

(Zeise's Salt)

$$K_2 PtCl_4 + C_2 H_4 + H_2 O \xrightarrow{\text{SnCl}_2} K[PtCl_3 (C_2 H_4)] \cdot H_2 O + KCl$$

Submitted by P. B. CHOCK,* J. HALPERN,* and F. E. PAULIK*
Checked by SAUL I. SHUPACK† and THOMAS P. DeANGELIS†

The original method[1] of preparation of Zeise's salt, $K[Pt(C_2 H_4)Cl_3] \cdot H_2 O$, and subsequent modifications thereof[2-4] all require either prolonged reaction times (7–14 days) or the use of high pressures. Furthermore, these procedures tend to yield products contaminated with potassium chloride and unreacted potassium tetrachloroplatinate(II). The improved procedure described below, which utilizes tin(II) chloride to catalyze the reaction between ethylene and the tetrachloroplatinate(II),[5,6] results in the formation of Zeise's salt of high purity and in high yield within a few hours at atmospheric pressure.

Procedure

To 45 ml. of 5 *M* aqueous hydrochloric acid in a 125-ml. Erlenmeyer flask is added 4.5 g. of potassium tetrachloroplatinate(II) (0.00108 mole). The flask is sealed with a rubber, serum cap and deoxygenated immediately by flushing for 30 minutes with nitrogen or ethylene through a polyethylene tube extending into the solution and attached to a needle inlet, with another needle as gas outlet. (Some undissolved potassium tetrachloroplatinate(II) may remain at this stage.) Forty milli-

*University of Chicago, Chicago, Ill. 60637.
†Villanova University, Villanova, Pa. 19085.

grams of hydrated tin(II) chloride, $SnCl_2 \cdot 2H_2O$ (0.0002 mole)* is placed in a 5-ml. flask which is sealed with a serum cap and deoxygenated by flushing with pure nitrogen with needles as gas inlet and outlet. With a hypodermic syringe, 5 ml. of deoxygenated distilled water is added to the tin(II) chloride, and the resulting suspension is transferred, also by means of a hypodermic syringe, to the flask containing the chloroplatinate(II). A stream of ethylene is bubbled slowly through the resulting reaction mixture, which is shaken periodically. During the course of 2–4 hours, the initially red-brown suspension turns yellow, and most of the solid dissolves as reaction proceeds. The reaction mixture is warmed to 40–45°C. and clarified by filtering through a sintered-glass filter (do not use paper). Cooling the filtrate in an ice bath yields a yellow precipitate of needle-shaped crystals of Zeise's salt, $K[PtCl_3(C_2H_4)] \cdot H_2O$, which is separated by filtration, washed with a small amount of ice water, and air-dried at room temperature. The yield is 3.6 g. (86%).† (Prolonged refrigeration of the mother liquor yields some additional product.) The infrared and visible-ultraviolet spectra (λ_{max}, 333 nm.; ϵ_{max}, 230) of this product (which is unaffected by further recrystallizations from 5 M HCl) are in excellent accord with literature data.[7,8]

Pumping *in vacuo* for 16 hours results in removal of the water of hydration, which yields $K[PtCl_3(C_2H_4)]$. *Anal.* Calcd. for $C_2H_4Cl_3KPt$: C, 6.55; H, 1.09; Cl, 29.0. Found: C, 6.65; H, 1.09; Cl, 28.52.‡

---

*Although the amount of tin(II) chloride used, and thereby the rate of the subsequent reaction, can be increased, the use of higher levels may be detrimental to the purity of the product and is not recommended.

†The checkers report that initial cooling in Dry Ice resulted in the precipitation of only 2 g. of Zeise's salt. Evaporating the filtrate and adding just enough methanol (*ca.* 10–15 ml.) to dissolve the solids, filtering off the KCl and other impurities such as tin(II) chloride and unreacted K_2PtCl_4, followed by rapid evaporation of the methanol, yielded a further 1.5 g. of Zeise's salt. Total yield, 85%.

‡The checkers report that, using essentially the same method, they were able to prepare the analogous *cis*-2-butene platinum complex in 70% yield.

Properties

Zeise's salt is obtained as well-formed, yellow, needle-shaped crystals. The compound is stable in the solid state at room temperature and decomposes with loss of ethylene at about 180°C. The chemical, physical, and structural properties have been characterized thoroughly and are described in the literature.[9-13]

References

1. W. C. Zeise, *Mag. Pharm.*, **35**, 105 (1830).
2. I. I. Chernyaev and A. D. Hel'man, *Ann. Secteur Platine, Inst. Chim. gen. (U.S.S.R.)*, **14**, 77 (1937).
3. J. Chatt and M. L. Searle, *Inorganic Syntheses*, **5**, 210 (1957).
4. W. MacNevin, A. Giddings, and A. Foris, *Chem. & Ind. (London)*, **1958**, 577.
5. R. Cramer, E. L. Jenner, R. V. Lindsey, and U. G. Stolbert, *J. Amer. Chem. Soc.*, **85**, 1691 (1963).
6. R. Pietropaolo, M. Graziani, and U. Belluco, *Inorg. Chem.*, **8**, 1506 (1969).
7. M. J. Grogan and K. Nakamoto, *J. Amer. Chem. Soc.*, **88**, 5454 (1966).
8. J. R. Joy and M. Orchin, *ibid.*, **81**, 305 (1959).
9. D. B. Powell and N. Sheppard, *Spectrochim. Acta*, **13**, 69 (1958).
10. J. W. Moore, *Acta Chem. Scand.*, **20**, 1154 (1966).
11. J. A. J. Jarvis, B. T. Kilbourn, and P. G. Owston, *Acta Cryst.*, **B27**, 366 (1971).
12. B. B. Bokii and G. A. Kukina, *Krystallografiya*, **2**, 3 (1967).
13. S. Maricic, C. R. Redpath, and J. A. Smith, *J. Chem. Soc.*, **1963**, 4905.

18. CHLOROBIS(CYCLOOCTENE)RHODIUM(I) AND –IRIDIUM(I) COMPLEXES

Submitted by A. van der ENT* and A. L. ONDERDELINDEN*
Checked by ROBERT A. SCHUNN†

The cyclooctene compounds $[MCl(C_8H_{14})_2]_n$, with M = Rh or Ir, are important starting materials for the preparation of rhodium(I)

*Unilever Research, Olivier van Noortlaan 120, Vlaardingen, The Netherlands.
†Central Research Department, E. I. du Pont de Nemours & Company, Wilmington, Del. 19898.

and iridium(I) complexes.[1,2] The compound $[RhCl(C_8H_{14})_2]_n$ can be separated in varying yields (35–60%) from solutions of rhodium(III) chloride 3-hydrate and cyclooctene in ethanol[3] after standing 3–5 days. Di-μ-chlorobis[bis(cyclooctene)iridium] can be prepared in 40% yield by refluxing chloroiridic(IV) acid and cyclooctene in 2-propanol.[4] The resulting product is always contaminated with an iridium hydride complex. The following modifications give better yields (70–80%) and an iridium(I) complex of higher purity.

A. CHLOROBIS(CYCLOOCTENE)RHODIUM(I)

$$RhCl_3 + 2C_8H_{14} + CH_3CH(OH)CH_3 \longrightarrow$$
$$RhCl(C_8H_{14})_2 + CH_3COCH_3 + 2HCl$$

Procedure

In a 100-ml., three-necked, round-bottomed flask, 2 g. (7.7 mmoles) of rhodium(III) chloride 3-hydrate is dissolved in an oxygen-free mixture of 40 ml. of 2-propanol and 10 ml. of water. Cyclooctene (6 ml.) is added. The solution is stirred for about 15 minutes under nitrogen. The flask is then closed and allowed to stand at room temperature for 5 days. The resulting reddish-brown crystals are collected on a filter, washed with ethanol, dried under vacuum, and stored under nitrogen at $-5°C$. The yield is 2.0 g. (74%). *Anal*. Calcd. for $RhC_{16}H_{28}Cl$: Rh, 28.72; C, 53.56; H, 7.81; Cl, 9.91. Found: Rh, 28.55; C, 53.76; H, 7.89; Cl, 9.76.

Properties

The solubility of $[RhCl(C_8H_{14})_2]_n$ in benzene and chloroform is too low for molecular-weight measurements. Its reddish-brown color darkens slowly in air.

B.   DI-μ-CHLOROBIS[BIS-(CYCLOOCTENE)IRIDIUM]

$$2(NH_4)_3 IrCl_6 + 4C_8H_{14} + 2CH_3 CH(OH)CH_3 \longrightarrow$$
$$[IrCl(C_8H_{14})_2]_2 + 6NH_4 Cl + 2CH_3 COCH_3 + 4HCl$$

Procedure

In a 250-ml., three-necked, round-bottomed flask, fitted with a nitrogen inlet and a reflux condenser, 6 g. (0.01 mole) of ammonium hexachloroiridate(III)* (43.1% Ir) is suspended in an oxygen-free mixture of 30 ml. of 2-propanol and 90 ml. of water.† Cyclooctene (12 ml.) is added. The mixture is refluxed on a water bath under a slow stream of nitrogen and with vigorous stirring for 3–4 hours. After cooling, the alcohol-water mixture is decanted, the last few milliliters being pipetted off. The orange oil remaining in the flask is allowed to crystallize under ethanol at 0°C. The yellow crystals are collected on a filter, washed with cold ethanol, dried under vacuum, and stored under nitrogen at room temperature. The yield is 4.7 g. (80%).‡ *Anal.* Calcd. for $Ir_2 C_{32} H_{56} Cl_2$: C, 42.89; H, 6.25; Cl, 7.93. Found: C, 43.12; H, 5.97; Cl, 7.84.

Properties

The results of molecular-weight measurements on a freshly prepared solution in benzene suggest a dimeric structure (found: M = 886; calcd., M = 895). In the solid state, $[IrCl(C_8H_{14})_2]_2$ decomposes slowly under the influence of atmospheric moisture. The compound is moderately soluble in benzene, chloroform, and carbon tetrachloride, but in general, these solutions are unstable for long periods of time. In comparison with the corresponding

---

*Available from Johnson, Matthey Company, Ltd., London, England.
†Similar results are obtained by using sodium or potassium chloroiridate(III).
‡Checkers found a yield of 74%.

rhodium complex, this compound is more reactive in oxidative addition reactions. This is demonstrated by the formation of iridium hydrides during reaction with hydrogen and hydrogen chloride, respectively.

References

1. S. Montelatici, A. van der Ent, J. A. Osborn, and G. Wilkinson, *J. Chem. Soc. (A)*, 1968, 1054.
2. M. A. Bennett and D. L. Milner, *J. Amer. Chem. Soc.*, 91, 6983 (1969).
3. L. Porri, A. Lionetti, G. Allegra, and A. Immirzi, *Chem. Commun.*, 1965, 336.
4. B. L. Shaw and E. Singleton, *J. Chem. Soc. (A)*, 1967, 1683.

19. COPPER(I) AMMONIUM TETRATHIOMOLYBDATE(VI)

$$MoS_4{}^{2-} + Cu^+ + NH_4{}^+ \longrightarrow CuNH_4MoS_4$$

Submitted by M. J. REDMAN*
Checked by ROSS H. PLOVNICK†

When hydrogen sulfide is passed into a strongly ammoniacal solution of copper(II) and molybdate(VI) ions, the initial copper-(II) sulfide precipitate is redissolved, the copper is reduced to the copper(I) state by sulfide ion, and the compound $CuNH_4MoS_4$, copper(I) ammonium thiomolybdate(VI), precipitates. The compound has been reported previously by Debray,[1] who was unable to characterize it. The following method of preparation is based upon Debray's work and gives superior yields and higher purity.

Procedure

The procedure should be carried out in an efficient fume hood. A solution of ammonium hydrogen sulfide is prepared by saturating a concentrated ammonium hydroxide solution with

*Ledgemont Laboratory, Kennecott Copper Corporation, Lexington, Mass. 02170.
†Cornell University, Clark Hall, Ithaca, N.Y. 14850.

hydrogen sulfide. To 150 ml. of this stirred solution (containing 23–26 g. of hydrogen sulfide as determined from weighing the ammonia solution before and after treatment) is added a solution consisting of 75 ml. of concentrated ammonia, 75 ml. of distilled water, 20 g. (0.016 mole) of reagent-grade hexaammonium hepta-molybdate (paramolybdate), $(NH_4)_6Mo_7O_{24} \cdot 4H_2O$, and 15 g. of reagent-grade copper(II) sulfate 5-hydrate, $CuSO_4 \cdot 5H_2O$ (0.06 mole). Alternatively, the sulfide solution can be added to the molybdate solution without any loss of yield.

A black precipitate appears which dissolves almost immediately to form a deep-red solution. The solution is boiled with stirring for 2–4 minutes, during which time a crystalline precipitate forms, which appears green in reflected light and purplish-red in transmitted light. The precipitate is filtered under suction and washed with 30 ml. of concentrated ammonia followed by 30 ml. of water and finally with 30 ml. of ethanol. It is then allowed to dry in air and finally in a desiccator. Yields of product range from 14.4 to 17.6 g. (79–96%).

Alternatively, the compound can be obtained by letting the deep-red solution, obtained after mixing the solutions, stand at room temperature for 2–4 hours. In this case the product is usually contaminated with crystals of ammonium tetrathio-molybdate(VI), $(NH_4)_2MoS_4$. The compound also can be obtained in poor yields by dissolving freshly precipitated copper(II) sulfide in ammoniacal ammonium molybdate solution saturated with hydrogen sulfide. Undissolved copper(II) sulfide is removed by filtration after 3–5 minutes, and the copper(I) ammonium thiomolybdate(VI) is obtained by boiling as above.

Copper(I) ammonium tetrathiomolybdate(VI) is diamagnetic, which suggests that copper(I) rather than copper(II) is present. A typical analysis gives a composition of $Cu(NH_4)_{1.07}Mo_{1.05}S_{4.09}$. *Anal.* Calcd. for $CuNH_4MoS_4$: Cu, 20.79; Mo, 31.38; S, 41.97; NH_4, 5.89. Found: Cu, 19.80; Mo, 31.39; S, 40.78; NH_4 (determined as NH_3), 5.98. The remaining material (2%) is possibly retained moisture.

Properties

The deep green crystals are converted to a red powder on grinding. The material is unaffected by air, but with water it decomposes slowly to form soluble ammonium thiomolybdate and insoluble copper(I) sulfide. The compound crystallizes in the tetragonal space group $I\bar{4}$ with unit-cell dimensions of $a = 8.000 \pm 0.004$ Å. and $c = 5.409 \pm 0.003$ Å.[2] These lattice constants were obtained by a method of least squares with data from x-ray diffraction patterns taken with a Norelco 114.6-mm.-diam. camera at 25°C. with $K\alpha$ cobalt radiation [$\lambda(K\alpha_1)1.7889$ Å., $\lambda(K\alpha_2)1.7928$ Å.].

References

1. M. Debray, *Compt. Rend.*, **96**, 1616 (1883).
2. W. P. Binnie, M. J. Redman, and W. J. Mallio, *Inorg. Chem.*, **9**, 1449 (1970).

Chapter Four

SOME SIGNIFICANT SOLIDS

In recent years the area of solid-state chemistry has become increasingly important to chemists. The development of memory cores for computers, phosphors, transistors, lasers, etc., has resulted in the synthesis of many new materials. Methods have had to be developed for both the preparation and characterization of these compounds.

In the past, most solids were prepared on a large scale by standard ceramic techniques, in which accurate control of the composition, as well as uniform homogeneity of the product, were not readily achieved. Unfortunately, this has sometimes led to uncertainty in the interpretation of the physical measurements. In recent years more novel methods have been developed to facilitate the reaction between solids. This is particularly true for the preparation of polycrystalline samples, on which the most measurements have been made. It is of utmost importance to prepare pure single-phase compounds, and this may be very difficult to attain. Even for a well-established reaction, careful control of the exact conditions is essential to ensure reproducible results. For any particular experiment, it is essential to devise a set of analytical criteria to which each specimen must be subjected. It will be seen from the solid-state syntheses included in this volume that one or more of the following common tests of "purity" are used to characterize a product.

1. Chemical analysis for cation composition, cation/anion ratio, and chemical impurities.

2. Crystallographic investigation, including the determination of lattice parameters, absence of foreign phases by x-rays, density determination, and optical examination.

3. Electrical conductivity (a.c. and d.c.), Hall measurements, and determination of Seebeck voltage.

4. Check of known magnetic properties, e.g., magnetic moment, Curie temperature determination, etc.

5. Optical properties, e.g., transmission, fluorescence, phosphorescence, and photoconductivity.

Obviously, it would be impractical to perform all the above tests on each new compound synthesized. The selection of appropriate characterization techniques is dependent upon the nature and properties of the compound being studied.

Most methods for synthesizing polycrystalline samples are basically similar and depend upon well-known ceramic techniques. The reactants are weighed out carefully, mixed, and placed in a suitable container. They are heated at a temperature high enough to initiate solid-solid reactions. This process is repeated several times until there is evidence of complete reaction. Among the many factors which must be controlled carefully are an accurate control of purity and composition of starting materials, an avoidance of impurities during the mixing process, a careful control of the atmosphere in the furnace to prevent oxidation or reduction, the firing temperatures and times, and the cooling rates. At present, the exact procedure for any particular compound must be carefully worked out and followed by analysis of the product.

For some investigations single crystals are essential, and for many they are desirable, both to determine the effects of anisotropy and to obtain higher purity. If the compound to be prepared melts congruently, large single crystals can generally be obtained by slow cooling of, or pulling from, the melt. Another common and simple method of growing crystals is from solution, either by evaporation or slow cooling of a molten flux. So far, most work with fluxes has been based on the slow-cooling method of crystallization, but there seems to be no reason against using controlled evaporation of the solvent at constant temperature, as is done for aqueous solutions.

In recent years, more novel methods have been developed by chemists in order to hasten the reaction between solids or to grow single crystals of new and exotic solids. Among those included in the following syntheses are electrolysis of fused salts, chemical transport, and hydrothermal crystal growth.

I. HALIDES AND OXYHALIDES

20. IRON(II) HALIDES

Submitted by G. WINTER*
Checked by D. W. THOMPSON† and J. R. LOEHE†

Iron(II) chloride and bromide may be obtained from the reaction of the metal with the appropriate hydrogen halide at elevated temperatures.[1] The chloride has also been made by the reduction of iron(III) chloride with hydrogen,[1] from iron(III) chloride and the metal in tetrahydrofuran,[2] and by the reaction of iron(III) chloride with chlorobenzene.[3] The iodide has been prepared from the metal and iodine in a sealed tube at elevated temperatures.[1]

The simplest procedure, dissolution of metallic iron in the aqueous mineral acid, suffers from the risk of accidental oxidation. The following relatively simple procedure overcomes this difficulty. This method, with minor modifications, has also been used successfully by the author for the preparation of chromium-(II) halides.

*Division of Mineral Chemistry, CSIRO, P.O. Box 124, Port Melbourne, Victoria 3207, Australia.
†College of William and Mary, Williamsburg, Va. 23185.

Procedure

A. IRON(II) HALIDES

$$Fe + 2HCl \longrightarrow FeCl_2 + H_2$$
$$Fe + 2HBr \longrightarrow FeBr_2 + H_2$$
$$Fe + 2HI \longrightarrow FeI_2 + H_2$$

The apparatus shown in Fig. 10 is flushed with nitrogen, and vessel B is stoppered. With the plug removed, a stream of nitrogen is passed through the sidearm and so provides an efficient gas curtain to prevent diffusion of atmospheric oxygen. Then 10 g. (0.18 g. atom) of iron powder (hydrogen reduced) is placed in vessel A, followed by 100 ml. of methanol* and the mineral acid, see Table I. The ensuing reaction is maintained under a nitrogen atmosphere at a vigorous rate by immersion of vessel A in hot water. It is completed within 2–3 hours. Cessation of hydrogen evolution must be ensured to prevent pressure from developing during later stages when the apparatus is closed.

---

*Methanol is used in preference to water to facilitate evaporation. Deaeration is not necessary.

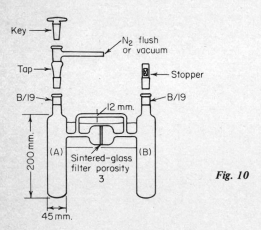

Fig. 10

TABLE I

Acid	Approx. strength, %	Amount, cc.	Approx. yield of FeX$_2$, g.	X, % Calcd.	X, % Found
HCl	32	40	21	56.5	56.2
HBr	50	45	40	74.1	73.7
HI	55	60	45	81.9	81.0*

*The checkers report a value of 78.2% I present in their samples of FeI$_2$.

The stream of nitrogen is discontinued and the apparatus closed by replacing the key in the tap. The greenish-gray solution is filtered through the sintered-glass disk into vessel B, and the filtrate cooled by immersion in a Dry Ice–alcohol slush bath. With the tap in the open position, the apparatus is evacuated through the sidearm by an efficient mechanical pump. The apparatus is sealed by closing the tap, and the solvent is evaporated from vessel B and condensed in vessel A by immersing A in the Dry Ice slush bath and slowly warming B to 100°C. When all the green crystalline hexamethanol solvate is converted into the white dimethanol solvate (about 3 hours), nitrogen is admitted. With the key of the tap removed and a stream of nitrogen flowing through the sidearm, the solvent collected in vessel A is poured out and discarded. The final desolvation is accomplished by heating the product in vessel B for 4 hours at 160°C. under vacuum. (For the iodide, heating for 2 hours at 100°C. is sufficient; excessive heating may result in loss of iodine.) This is conveniently achieved by connecting the apparatus to a mechanical pump protected by a suitable vapor trap, and immersing vessel B in an oil bath or small furnace. After cooling, nitrogen is again admitted, and the product is transferred to storage containers.

The amounts of mineral acid recommended for 10 g. of iron powder, the approximate yield of product, and halogen analyses are shown in Table I.

B. IRON(II) IODIDE

$$Fe + I_2 \longrightarrow FeI_2$$

For the preparation of FeI_2 an alternative procedure involving reaction of the elements may be used.

The apparatus (Fig. 10) is flushed with nitrogen and protected from ingress of oxygen as described in Procedure A. Five grams (0.09 g. atom) of iron powder (hydrogen reduced) is placed in vessel A, and then 100 ml. of methanol and 2 ml. of 55% hydroiodic acid to activate the iron. Finely powdered, sublimed iodine, 23 g. (0.09 mole), is added over a period of $\frac{1}{2}$ hour, the methanol being maintained at its boiling point by occasional immersion of vessel A in hot water. The dark green solution is filtered through the sintered-glass disk into vessel B. The evaporation of the solvent and desolvation of the product are performed as described in Procedure A. Approximate yield is 23 g. *Anal.* Calcd. for FeI_2: I, 81.9. Found: I, 81.5.

Properties

Iron(II) chloride is a pale buff-colored solid, iron(II) bromide is pale yellow, and iron(II) iodide is deep red. They can be stored in air, provided moisture is rigorously excluded.

References

1. G. Brauer, "Handbook of Preparative Inorganic Chemistry," Academic Press, Inc., New York, 1965.
2. M. F. Hawthorne, T. D. Andrews, P. M. Garrett, F. P. Olsen, M. Reintjes, F. N. Tebbe, L. F. Warren, P. A. Wegner, and D. C. Young, *Inorganic Syntheses,* **10,** 112 (1967).
3. P. Kovacic and N. O. Brace, *Inorganic Syntheses,* **6,** 172 (1960).

21. NIOBIUM(IV) FLUORIDE AND NIOBIUM(V) FLUORIDE

$$5SnF_2 + 2Nb \longrightarrow 5Sn + 2NbF_5$$
$$4NbF_5 + Si \longrightarrow 4NbF_4 + SiF_4$$

Submitted by FRANK P. GORTSEMA*
Checked by JAMES B. BEAL, JR.,† and KARL SCHMIDT†

The compound niobium tetrafluoride is of importance because of the increased interest in reactions of metal tetrafluorides with basic ligands. Previous syntheses given for this compound are time consuming, generally cannot be scaled up to prepare larger quantities, or require expensive equipment not always readily available. In this synthesis, simple methods are described for the preparation of niobium(V) fluoride and niobium(IV) fluoride, which should make these compounds readily available to chemists for further comparison studies with other metal tetrafluorides.

The following preparative method (of general utility) includes the reduction of niobium(V) fluoride with elements such as silicon, phosphorus, and boron,[1] which yield highly volatile fluorides. Silicon reductions have been studied in greatest detail. Niobium(IV) fluoride can be prepared in good yield by the reaction described by the equation:

$$4NbF_5 + Si \longrightarrow 4NbF_4 + SiF_4$$

*Union Carbide Research Institute, Tarrytown, N.Y. 10591.
†Ozark-Mahoning Company, 1870 S. Boulder Ave., Tulsa, Okla. 74119 (J. B. Beal is presently at the University of Montevallo, Montevallo, Ala.).

A. NIOBIUM(V) FLUORIDE

Procedure

Powdered niobium metal, 20.0 g. (−200 mesh),* and tin(II) fluoride, 52.0 g. (40 mesh),† are mixed in a molybdenum crucible in an Inconel- or nickel-pipe reactor approximately 3 in. in diameter and 10 in. long and heated to 400–500°C. in a stream of dry nitrogen. The niobium(V) fluoride volatilizes from the reaction mixture and condenses on the water-cooled lid of the reactor, which leaves metallic tin in the crucible. The yield of niobium(V) fluoride is 21.1 g., or 95% of theoretical. A very small amount of blue niobium oxyfluoride (composition of variable oxygen and fluorine content) often forms as an impurity because of the presence of minute amounts of oxygen. *Anal.* Calcd. for NbF_5 : Nb, 49.44; F, 50.56. Found: Nb, 49.43; F, 50.2.

Properties

The white crystalline solid is very hygroscopic and has an m.p. of 80°C. and a b.p. of 235°C.

B. NIOBIUM(IV) FLUORIDE

Procedure

If niobium(V) fluoride is purchased from commercial vendors,‡ it must be purified by sublimation before use. Charges of niobium(V) fluoride (5–50 g. in a platinum crucible) are heated in

---

*Niobium (−200 mesh), 99.8+% Nb; Ta, <500 p.p.m.; O, <200 p.p.m.; Fansteel Metallurgical Company, Chicago, Ill. 60600.

†Tin(II) fluoride (40 mesh), City Chemical Company, 130 W. 22d St., New York, N.Y. 10011.

‡Ozark-Mahoning Company, 1870 S. Boulder Ave., Tulsa, Okla. 74119.

a glass or silica sublimation apparatus at 5-20 torr and 50-100°C. The sublimate collects on a water-cooled cold finger. All transfer operations involving niobium(V) fluoride are carried out in a glove box in an atmosphere of dry nitrogen.

The preparation of niobium(IV) fluoride is conducted in the reactor shown in Fig. 11, which is constructed from nickel or copper pipe and tubing, 1/2 in. in diameter by 8 in. long, with silver-soldered or Swagelok* connectors. The most expensive part of the apparatus is the large Monel valve.† A stainless-steel or brass valve can be substituted, but corrosion of the valve seat does occur. Since the Swagelok fitting A on the valve gets very hot because of its proximity to the furnace, the threads should be coated with a high-temperature lubricant such as Silver Goop* to prevent metal seizure. A pressure gage may be connected to one arm of the apparatus, B.

*Crawford Fitting Company, 29500 Solon Rd., Solon, Ohio 44139.
†Whitey Research Tool Company, Emeryville, Calif. 94608.

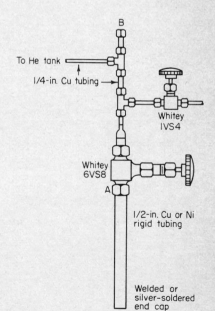

Fig. 11. Metal reactor for the preparation of niobium(IV) fluoride.

In a typical experiment, 4.63 g. (0.025 mole) of niobium(V) fluoride is mixed with 0.348 g. (0.012 mole) of −200-mesh silicon and loaded into the reactor inside a glove box under dry nitrogen. The assembled reactor is connected to a helium tank and repeatedly pressurized and depressurized to rid the system of air. The entire system is pressurized to 50 p.s.i., the Monel reactor valve is closed, and the reaction chamber is placed in a vertical furnace at 350°C. The pressure in the external lines is left at 50 p.s.i. to prevent entry of air.

The loaded reactor is heated for 24 hours, then removed from the furnace, and cooled to room temperature. The system is depressurized and the silicon tetrafluoride expelled into a hood, or bled into a water scrubber.* The system is purged several times with helium and the heating procedure repeated for an additional 4 hours, after which the silicon(IV) fluoride removal procedure is repeated as well. The solid black product is niobium(IV) fluoride, 3.752 g. or 90% of theoretical yield. *Anal.* Calcd. for NbF_4: Nb, 55.0; F, 45.0. Found: Nb, 55.1; F, 44.3.

For larger quantities the above procedure is used with a greater number of silicon(IV) fluoride elimination steps. Fifty-gram batches have been prepared by scaling up this technique and using a larger metal-pipe reactor.

Properties

Niobium(IV) fluoride is a black, very hygroscopic powder. It reacts with moisture in the air to give a viscous brown liquid. Complete hydrolysis results in white NbO_2F. Niobium(IV) fluoride reacts vigorously with water to give a brown solution which forms a brown uncharacterized precipitate.

---

*Silicon(IV) fluoride can be scrubbed by passing the gas into a vessel containing water or aqueous ammonia. The silicon(IV) fluoride reacts with water to produce fluorosilicic acid [dihydrogen hexafluorosilicate (2–)] by the reaction shown by the equation

$$2SiF_4(g) + 2H_2O \longrightarrow SiO_2(s) + 2H^+ + SiF_6^{--} + 2HF(g)$$

Powder x-ray diffraction shows the crystal structure of niobium-(IV) fluoride to be tetragonal, $D_{4h}^{17} - I\,4/mmm$ with cell constants $a_0 = 4.081$ A. and $c_0 = 8.162$ A., $\rho_{obs.} = 4.01$ g./cc., $\rho_{calc.} = 4.13$ g./cc.[2]

When niobium(IV) fluoride is heated to approximately $400°$C., it begins to lose niobium(V) fluoride. At $400°$C. cubic Nb_6F_{15} is formed according to the reaction described by the equation:

$$5NbF_4 \longrightarrow 2NbF_{2.5} + 3NbF_5$$

When the decomposition reaction is performed in silica or glass containers, $Nb(O,F)_3$* results. These materials are gray-black to black, refractory materials having the rhenium trioxide type of cubic structure, with a cell constant varying from 3.889 to 3.917 A.[2] These materials contain variable amounts of oxygen and have often been erroneously identified as NbF_3.

References

1. F. P. Gortsema and R. Didchenko, assigned to Union Carbide Corporation, U.S. patent 3,272,592,
2. F. P. Gortsema and R. Didchenko, *Inorg. Chem.*, 4,182 (1965).
3. H. Schäfer, H. G. Schnering, K. J. Niehues, and H. G. Nieder-Vahrenholz, *J. Less-Common Metals*, 9, 95 (1965).

22. TUNGSTEN OXYHALIDES

Submitted by J. TILLACK†
Checked by T. M. BROWN‡ and H. SCHÄFER §

A number of pure tungsten oxyhalides (WO_2Cl_2, $WOCl_4$, $WOCl_3$, $WOCl_2$, WO_2Br_2, $WOBr_4$, $WOBr_3$, $WOBr_2$, and WO_2I_2) have been prepared by chemical transport techniques, which yield very pure

*The oxyfluoride has a variable composition with respect to oxygen and fluorine.

†Philips Zentrallaboratorium GmbH, Laboratorium Aachen, Germany.
‡Department of Chemistry, Arizona State University, Tempe, Ariz. 85281.
§ Westfalische Wilhelms, Universität Anorganisch Chemisches Institut, 44 Münster, Germany.

compounds with a minimum of experimental effort. The method has been successfully used for many years, particularly in the laboratory of Harald Schäfer at the University of Münster, for the preparation of many halides and oxyhalides of the transition elements.[1]

A. TUNGSTEN(VI) DICHLORIDE DIOXIDE

$$2WO_3 + WCl_6 = 3WO_2Cl_2$$

Procedure

A Pyrex glass tube (15 cm. in length, 2.4 cm. in diameter) is filled with a stoichiometric mixture of 9.274 g. (40 mmoles) of tungsten(VI) oxide and 7.931 g. (20 mmoles) of tungsten(VI) chloride (prepared from tungsten and chlorine). To this charge a small excess of 1 mg. WCl_6/ml. of tube volume is added. The tube, with one end cooled in liquid nitrogen, is attached to a vacuum system and evacuated to 10^{-2} torr (Fig. 12). The tube is sealed off at A and located in a temperature gradient of 350/275°C. with the reactants placed in the hotter part of the tube (Fig. 13).

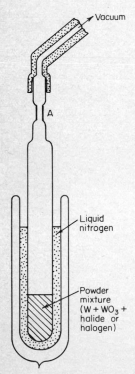

The two-zone furnace used to produce the temperature gradient T_2/T_1 consists of two, hollow, alumina cylinders (8 cm. o.d., 4 cm. i.d., 35 cm. length), independently wound with Kanthal wire. These cylinders are mounted inside an iron frame which is filled with asbestos wool and covered with asbestos board. The furnace is tilted slightly (approximately 10°) to prevent the liquid products from flowing back into the hotter part of the tube and causing a possible explosion.

Fig. 12. Filling of the reaction tube.

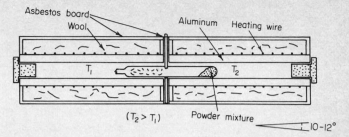

Fig. 13. Heating equipment.

After about 20 hours the reaction is complete, and the growth zone is filled with yellow flakes of tungsten(VI) dichloride dioxide crystals. Then while it is still hot, the tube is placed in a reverse temperature gradient of 200°C./room temperature, so that the more volatile impurities (for example, WCl_6 and $WOCl_4$) are condensed in the empty (cold) end of the reaction tube. The yield is approximately 17 g. of tungsten(VI) dichloride dioxide (98% based on the amount of tungsten(VI) oxide used). If further purification or better crystallization is desired, the crystals may be sublimed as often as desired in the presence of a slight excess of tungsten(VI) chloride (for example, 1 mg. of tungsten(VI) chloride per milliliter of tube volume). The analysis for tungsten(VI) dichloride dioxide gave the following results. *Anal.* Calcd. for WO_2Cl_2: W, 64.12; Cl, 24.72; O, 11.16. Found: W, 64.22; Cl, 24.67; O, 11.12.

Properties

The thin yellow flakes of tungsten(VI) dichloride dioxide, and the compounds to be mentioned later as well, are sensitive to atmospheric moisture. Therefore the reaction tubes must be opened in a dry nitrogen atmosphere and stored either under vacuum or in a dry inert gas. The detailed physical and chemical properties of tungsten(VI) dichloride dioxide are described in the literature.[1]

In all cases chemical analyses were made by the H-tube method.[2,3] The substance to be analyzed is weighed in a previously tared glass crucible, which is put into the shorter leg of the H tube. The other leg contains a solution of 1.5 ml. of concentrated nitric acid and 3.0 ml. of silver nitrate (100 mg. silver nitrate/ml. water). The H tube is sealed off and put into an oven at 120-140°C. for 12-24 hours until the reaction is complete. The crucible contains tungsten(VI) trioxide, which can be weighed after drying at 350-400°C. in the open air. The halogen is absorbed by the silver nitrate solution and so forms the corresponding silver halide, which can be determined readily in the conventional way. The percentages of tungsten and halogen are added, and the oxygen content of each sample is found by difference. The analyses for tungsten(VI) dichloride dioxide gave the following results. *Anal.* Calcd. for WO_2Cl_2: W, 64.12; Cl, 24.72; O, 11.16. Found: W, 64.22; Cl, 24.67; O, 11.12.

B. TUNGSTEN(VI) TETRACHLORIDE OXIDE

$$WO_3 + 2WCl_6 = 3WOCl_4$$

Procedure

A stoichiometric mixture of 3.478 g. (15 mmoles) of tungsten-(VI) oxide and 11.897 g. (30 mmoles) of tungsten(VI) chloride is placed in the reaction tube, and an excess of 1 mg. of tungsten(VI) chloride per milliliter of tube volume is added. The tube is sealed under vacuum and heated in a temperature gradient of 200/175°C. with the reaction mixture placed in the hotter part of the furnace (Fig. 14). After about 10 hours the reaction is complete, and 15 g. of tungsten(VI) tetrachloride oxide are obtained (98% yield). The analyses of the sublimed product by the H-tube method are somewhat inaccurate, because of the extreme moisture sensitivity of this substance (loss of hydrogen chloride by hydrolysis). *Anal.* Calcd. for $WOCl_4$: W, 53.82; Cl, 41.50; O, 4.68. Found: W, 54.20; Cl, 41.40; O, 4.46.

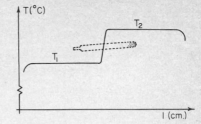

Fig. 14   Temperature distribution.

Properties

Tungsten(VI) tetrachloride oxide forms long red needles which are extremely sensitive to atmospheric moisture. It is relatively soluble in organic solvents and gives a dark-red solution in dioxane and in acetone. The latter becomes blue on standing. A dark green solution is formed in cyclohexanone. In ethanol and 2-methoxy-ethanol hydrogen chloride gas is evolved and a clear solution is formed. In ammonia or sodium hydroxide solution, tungsten(VI) tetrachloride oxide reacts vigorously and yields a colorless solution. It reacts with water, hydrochloric acid, and nitric acid to form a yellow-green precipitate.

C.   TUNGSTEN(V) TRICHLORIDE OXIDE

$$W + 2WO_3 + 3WCl_6 = 6WOCl_3$$

Procedure

A stoichiometric mixture of 1.834 g. (10 mg. atoms) of tungsten powder, 4.637 g. (20 mmoles) of tungsten(VI) oxide, and 11.897 g. (30 mmoles) of tungsten(VI) chloride is put into a reaction tube. An excess of 8 mg. of WCl_6 per milliliter of tube volume is added. The tube is sealed under vacuum and placed in a temperature gradient of 450/230°C.

After about 40 hours the reaction is complete, and the tube is placed immediately in a reverse temperature gradient of 200°C./ room temperature in order to separate the more volatile species from the desired product. The preparation of tungsten(V) trichloride oxide, first reported by Fowles and Frost,[4,5] is very dependent on the amount of excess tungsten(VI) chloride and the temperature gradient used. Employing the above conditions, 15.2 g. of tungsten(V) trichloride oxide (83% yield) was obtained.

Properties

Tungsten(V) trichloride oxide forms flexible black shiny needles, 10–20 mm. in length, which are somewhat hygroscopic and rather similar to tungsten(V) tribromide oxide. It is not attacked by cold, dilute solutions of hydrochloric, sulfuric, or nitric acid or of ammonia or organic solvents, such as 2-methoxyethanol, toluene, acetone, and ethanol. Hot water attacks tungsten(V) trichloride oxide quickly to form a dark blue solution and a precipitate. Cold solutions of sodium hydroxide or hot solutions of ammonia attack the compound rapidly, and the black needles acquire a light brown surface. Solutions of ammonia or sodium hydroxide containing 3% hydrogen peroxide attack the compound immediately, yielding a clear solution.

It has been determined by differential thermal analyses (D.T.A.) that tungsten(V) trichloride oxide decomposes above 290°C. without a sharp transition point. On cooling, a transition is observed at 210°C., the melting point of tungsten(VI) tetrachloride oxide. Therefore the decomposition of tungsten(V) trichloride oxide seems to take place by the following reaction:

$$2WOCl_3 (s) = WOCl_2 (s) + WOCl_4 (g)$$

The following analyses were made by the H-tube method. *Anal.* Calcd. for $WOCl_3$: W, 60.05; Cl, 34.74; O, 5.21. Found: W, 60.38;

Cl, 34.53; O, 5.09. The density of tungsten(V) trichloride oxide, found pycnometrically in toluene at $25.0 \pm 0.1°C.$, is 4.75 g./cc.

Note

The reaction times for these preparations depend considerably on the grain size of the tungsten powder used, the homogeneity of the mixture (W/WO_3), and the temperatures used.

D. TUNGSTEN(IV) DICHLORIDE OXIDE

$$W + WO_3 + WCl_6 = 3WOCl_2$$

Procedure

Tungsten(IV) dichloride oxide was first prepared both by Eliseev, Gluckhov, and Gaidaenko,[6] and by Tillack et al.[7] through the reduction of tungsten(VI) tetrachloride oxide with tin(II) chloride. In the present synthesis a stoichiometric mixture of 3.677 g. (20 mg. atoms) of tungsten powder, 4.637 g. (20 mmoles) of tungsten(VI) oxide, and 7.931 g. (20 mmoles) of tungsten(VI) chloride is put into the reaction tube. An excess of 2–3 mg. of tungsten(VI) chloride per milliliter of tube volume is added. In this case, when a temperature gradient of $450/250°C.$ is employed, approximately 15 g. of product is obtained (92% yield). *Anal.* Calcd. for $WOCl_2$: W, 67.91; Cl, 26.19; O, 5.90. Found: W, 68.13; Cl, 26.13; O, 5.74.

The crude product usually contains excess chlorine $(WOCl_{2.3})$, which can be removed by chemical transport in the presence of 2 mg. tungsten(VI) chloride per milliliter of tube volume in a temperature gradient of $520/250°C.$ The proposed chemical transport equation is:

$$WOCl_2 (s) + 2WCl_6 (g) = WOCl_4 (g) + 2WCl_5 (g)$$

Properties

The properties, crystal habit, and x-ray pattern of tungsten(IV) dichloride oxide are very similar to those of molybdenum(IV) dichloride oxide.[8] Stoichiometric tungsten(IV) dichloride oxide, which forms gold-brown needles, is stable under atmospheric conditions and is not attacked by water, dilute or concentrated cold acids, ammonia, or organic solvents, such as acetone, ethanol, 2-methoxyethanol, chloroform, and diethyl ether. However, it decomposes in a solution of sodium hydroxide and forms a black precipitate, which disappears when hydrogen peroxide is added and yields a clear, yellow solution. The density of tungsten(IV) dichloride oxide, determined pycnometrically as previously mentioned, is 5.92 g./cc.

E. TUNGSTEN(VI) DIBROMIDE DIOXIDE

$$W + 2WO_3 + 3Br_2 = 3WO_2Br_2$$

Procedure

■ *Caution. When bromine or iodine is used, preliminary heating of the reactants is necessary to prevent an explosion of the tube because of the high partial pressure of the halogen. During this heating period only one end of the tube (containing the mixture of tungsten powder and tungsten(VI) oxide) is placed in the furnace to allow an initial reaction between the powdered mixture and the halogen until the free halogen is consumed and forms higher halides and oxyhalides of tungsten.*

A stoichiometric mixture of 2.758 g. (15 mg. atoms) of tungsten powder, 6.956 g. (30 mmoles) of tungsten(VI) oxide, and 7.192 g. (45 mmoles) of bromine is put into the reaction tube. A small excess of 2 mg. of bromine per milliliter of tube volume is added to prevent the formation of lower tungsten bromides or bromide oxides.

The tube is sealed under vacuum and placed in a temperature gradient of 400/40°C. until the bromine has completely reacted. After this preliminary firing, the reaction tube is placed in a temperature gradient of 450/325°C. for 15 hours. The crude product is purified by a second sublimation. The more volatile impurities, $WOBr_4$ and WBr_5, are separated in the usual way, which leaves behind pure crystalline tungsten(VI) dibromide dioxide. The product yield is 15 g. (97%). *Anal.* Calcd. for WO_2Br_2: W, 48.95; Br, 42.53; O, 8.52. Found: W, 48.78; Br, 41.58; O, 8.65.

Properties

Tungsten(VI) dibromide dioxide is sensitive to atmospheric moisture. It yields a clear, colorless solution in dilute and concentrated solutions of ammonia or sodium hydroxide. No reaction is observed with hot or cold concentrated hydrochloric acid, or dry organic solvents.

F. TUNGSTEN(VI) TETRABROMIDE OXIDE

$$2W + WO_3 + 6Br_2 = 3WOBr_4$$

Procedure

■ *Caution. Because of the danger of explosion, the reactants must undergo a preliminary heating in a temperature gradient of 400/40°C. until the bromine has completely reacted.*

The reaction tube is filled with a stoichiometric mixture of 3.677 g. (20 mg. atoms) of tungsten powder, 2.318 g. (10 mmoles) of tungsten(VI) oxide, and 9.589 g. (60 mmoles) of bromine. A small excess of 2 mg. of bromine per milliliter of tube volume is added. After the preliminary firing, the tube is placed in a temperature gradient of 425/250°C. for 15 hours.

The crude product contains a small amount of the less volatile tungsten(VI) dibromide dioxide as an impurity, which can be separated by sublimation at 120°C. under dynamic vacuum (10^{-3} torr). A yield of approximtely 15 g. (96%) of pure tungsten(VI) tetrabromide oxide is obtained after sublimation. Large crystals can be obtained by sublimation of the purified product in another tube with 1 mg. of bromine per milliliter of tube volume in a temperature gradient of 220/160°C. *Anal.* Calcd. for $WOBr_4$: W, 35.40; Br, 61.52; O, 3.08. Found: W, 35.37; Br, 61.56; O, 3.08.

Properties

Tungsten(VI) tetrabromide oxide is a dark-brown crystalline compound, which is obtained in the form of needles or flakes. It is extremely sensitive to atmospheric moisture, decomposes rapidly in water, which leaves a gray-green precipitate, and is completely dissolved in solutions of ammonia or sodium hydroxide, which form clear, colorless solutions. It is slightly soluble in concentrated hydrochloric acid, dioxane (yellow solution), 2-methoxyethanol (colorless solution), and acetone (green solution). It has a melting point of 322°C., as determined by D.T.A.

G. TUNGSTEN(V) TRIBROMIDE OXIDE

$$2W + WO_3 + 4.5Br_2 \longrightarrow 3WOBr_3$$

Procedure

■ *Caution. As described above, a preliminary firing of the reactants in a temperature gradient of 400/40°C. is necessary in order to eliminate the possibility of an explosion.*

A stoichiometric mixture of 3.677 g. (20 mg. atoms) of tungsten powder, 2.318 g. (10 mmoles) of tungsten(VI) oxide, and 7.192 g. (45 mmoles) of bromine is put into the reaction tube, and a small excess of 1 mg. of bromine per milliliter of tube volume is added. After the preliminary firing, the tube is placed in a temperature gradient of 450/350°C. for about 30 hours. The reaction tube is then turned around in the furnace in order to purify the crude product by reverse transport from the hotter to the colder end of the tube, supposedly according to the equation:

$$WOBr_3 (s) + 0.5 Br_2 = WOBr_4 (g)$$

Following this chemical transport, the tube is immediately placed in a temperature gradient of 250°C./room temperature in order to vaporize the more volatile species, $WO_2 Br_2$, $WOBr_4$, WBr_6. The yield is approximately 13 g. of pure tungsten(V) tribromide oxide (98%). *Anal.* Calcd. for $WOBr_3$: W, 41.83; Br, 54.53; O, 3.64. Found: W, 42.03; Br. 54.56; O, 3.39.

Properties

Tungsten(V) tribromide oxide, first prepared by Crouch, Fowles, Frost, Marshall, and Walton[5] by reducing tungsten tetrabromide oxide with aluminum powder, forms flexible blue-black needles, 10–20 mm. in length, which are stable in the atmosphere.[9] It is not attacked by hot or cold concentrated hydrochloric acid, acetic acid, sodium hydroxide, ammonia, or organic solvents, such as acetone, cyclohexanone, diethyl ether, petroleum ether, chloroform, benzene, and dioxane. Boiling water attacks $WOBr_3$ within a few seconds, yielding an intense blue solution and a dark, voluminous precipitate. It is similarly attacked by ethanol and 2-methoxyethanol. The rate of these last two reactions is accelerated by the addition of a solution of zinc sulfate. Tungsten(V) tribromide oxide reacts very little with weak oxidizing solutions (for example, 3% hydrogen peroxide, 2 N sulfuric acid, 2 N sodium hydroxide, or 2 N ammonia), but it is completely soluble in an alkaline solution of 30% hydrogen peroxide, which gives a colorless solution.

Differential thermal analysis has shown that tungsten(V) tribromide oxide decomposes above 420°C. without melting. The density, determined pycnometrically, is 5.87 g./cc.

H. TUNGSTEN(IV) DIBROMIDE OXIDE

$$2W + WO_3 + 3Br_2 = 3WOBr_2$$

Procedure

■ *Caution. To eliminate the possibility of explosion, a preliminary heating of the reactants in a temperature gradient of 450/40°C. is necessary.*

A stoichiometric mixture of 3.677 g. (20 mg. atoms) of tungsten powder, 2.318 g. (10 mmoles) of tungsten(VI) oxide, and 4.795 g. (30 mmoles) of bromine is placed in the reaction tube, and a small excess of 3–5 mg. of bromine per milliliter of tube volume is added. After the preliminary heating, the tube is placed in a temperature gradient of 580/450°C. with the charge mixture at the hotter end of the furnace. After about 100 hours the charge is completely transported to the cooler end of the furnace, and the reaction tube is reversed in order to purify the crude product. The equation for the transport reaction is:

$$WOBr_2\,(s) + Br_2\,(g) = WOBr_4\,(g)$$

Properties

Tungsten(VI) dibromide oxide, first prepared by Tillack and Kaiser,[10] forms well-crystallized, gray-black shiny needles. It is not attacked by water, dilute acids, ammonia, or organic solvents, such as ethanol, acetone, toluene, or trichloroethylene. Although the crystals appear unchanged in concentrated hydrochloric acid, a light-blue solution results. Concentrated nitric or sulfuric acid decomposes the compound, especially when heated. A warm solution of 2 N sodium hydroxide or ammonia containing 3% hydrogen peroxide attacks the compound immediately, yielding a clear, light-yellow solution. The density of tungsten(IV) dibromide oxide is 6.13 g./cc.

I. TUNGSTEN(VI) DIIODIDE DIOXIDE

$$W + 2WO_3 + 3I_2 = 3WO_2 I_2$$

Procedure

A stoichiometric mixture of 1.839 g. (10 mg. atoms) of tungsten powder, 4.637 g. (20 mmoles) of tungsten(VI) oxide, and 7.614 g. (30 mmoles) of iodine is put into the reaction tube, and an excess of 12 mg. of iodine per milliliter of tube volume is added to prevent the formation of tungsten(V) iodide dioxide or tungsten(IV) oxide. The tube is sealed under vacuum and placed in a temperature gradient of 100°C./room temperature until the iodine has sublimed to the empty half of the tube. The tube is placed in a temperature gradient of 525/200°C. for approximately 70 hours, and then the hot tube is turned around and immediately placed in a temperature gradient of 200°C./room temperature, so that the unused iodine is separated from the tungsten(VI) diiodide dioxide. (It has been observed that traces of moisture, which are normally present in the iodine used, accelerate the reaction.) If further purification of the product is desired, another sublimation is carried out with 10–12 mg. of iodine per milliliter of tube volume to prevent thermal decomposition. The normal yield of pure crystalline tungsten(VI) diiodide dioxide is 8 g. (60%) due to the kinetically controlled chemical transport reaction. *Anal.* Calcd. for $WO_2 I_2$: W, 39.15; I, 54.04; O, 6.81. Found: W, 39.27; I, 53.93; O, 6.80.

Properties

Tungsten(VI) diiodide dioxide is obtained as thin, shiny platelets of a dark brown or brassy color.[11-15] The crystals decompose slowly in the atmosphere and form a gray-green mixture, but they are stable in vacuum or dry, inert gases. The compound reacts slowly with water or dilute (2 N) solutions of hydrochloric, sulfuric, and nitric acids, forming a gray-green colored layer on the surface of the crystals. It is quickly attacked by hot dilute acid and hot or cold concentrated nitric acid, with the formation of nitrogen(IV) oxide and iodine. Colorless solutions are obtained with concentrated solutions of sodium hydroxide or ammonia, but no reaction is observed with concentrated hydrochloric acid or organic solvents.

References

1. H. Schäfer, "Chemical Transport Reactions," Academic Press, Inc., New York, 1964.
2. H. Schäfer and K. D. Dohmann, *Z. Anorg. Allgem. Chem.*, **300**, 1 (1959).
3. J. Tillack, *Z. Analyt. Chem.*, **239**, 81 (1968).
4. G. W. A. Fowles and J. T. Frost, *Chem. Commun.*, **9**, 252 (1966).
5. P. C. Crouch, G. W. A. Fowles, J. L. Frost, P. R. Marshall, and R. A. Walton, *J. Chem. Soc. (A)*, **1968**, 1061.
6. S. S. Eliseev, J. A. Gluckhov, and N. V. Gaidaenko, *Russ. J. Inorg. Chem. (English Transl.)*, **14**, 328 (1969).
7. J. Tillack, R. Kaiser, G. Fisher, and P. Eckerlin, *J. Less-Common Metals*, **20**(2), 171 (1969).
8. H. Schäfer and J. Tillack, *J. Less-Common Metals*, **6**, 152 (1964).
9. J. Tillack and R. Kaiser, *Angew. Chemie*, **80**, 286 (1968).
10. J. Tillack and R. Kaiser, *ibid.*, **81**, 141 (1969).
11. J. Tillack, P. Eckerlin, and J. H. Dettingmeijer, *Angew. Chemie*, **78**, 451 (1966).
12. J. H. Dettingmeijer and B. Meinders, (I) *Z. Anorg. Allgem. Chem.*, **357**, 1 (1968).
13. J. Tillack, (II) *ibid.*, **357**, 11 (1968).
14. H. Schäfer, D. Giegling, and K. Rinke, (III) *ibid.*, **357**, 25 (1968).
15. J. H. Dettingmeijer, J. Tillack, and H. Schäfer, (IV) *ibid.*, **369**, 161 (1969).

23. INDIUM(III) OXYFLUORIDE

(Indium(III) Oxide Fluoride)

$$In_2O_3 + InF_3 \xrightarrow{900°C.} 3InOF$$

Submitted by B. CHAMBERLAND*†
Checked by J. PORTIER‡

Attempts[1] to prepare indium(III) oxyfluoride by a displacement reaction of InOX (X = Cl, I) with fluorine led to decomposition under the conditions utilized. A hydrothermal reaction of $InF_3 \cdot 3H_2O$ and HF yielded[2] only $In(OH)F_2$. Dehnicke et al.[3] describes the thermal decomposition product of $In(ONO_2)_2F$ as InOF, but gives no characterization of the material. Mermant, Belinski, and Lalau-Keraly[4] have studied the thermal decomposition of $InF_3 \cdot 3H_2O$ but produced only indium(III) oxyfluoride contaminated with indium(III) oxide. Pure indium(III) oxyfluoride was isolated[5] by a variety of methods in which air, moisture, and HF were carefully excluded. The most convenient preparation[5,6] of pure indium(III) oxyfluoride is by a solid-state reaction involving the direct combination of indium(III) oxide with indium(III) fluoride. However, Hahn and Nickels[7] were unsuccessful in their efforts to prepare indium(III) oxyfluoride by this method because their reaction system was not entirely sealed, and their product was hydrolyzed by moisture to yield indium(III) oxide as the sole product.

*University of Connecticut, Storrs, Conn. 06268.

†Original synthesis carried out at Central Research Department, E. I. du Pont de Nemours & Company, Wilmington, Del. 19898.

‡Université de Bordeaux, Service de Chemie Minérale Structurale de la Faculté des Sciences de Bordeaux, 351, cours de la Liberation, 33, Talence, France.

This direct method of metal oxyfluoride synthesis can also be applied to the preparation of other compounds such as $FeOF$[8] and $TlOF$.[9,10]

Preparation of Indium(III) Oxyfluoride

The synthesis involves the direct reaction of stoichiometric amounts of anhydrous indium(III) fluoride and indium(III) oxide. Anhydrous indium(III) fluoride can be prepared[11] by the thermal decomposition of $(NH_4)_3[InF_6]$ in a stream of fluorine gas, or it can be purchased in high-purity form from Ozark-Mahoning Company.* Pure indium(III) oxide, In_2O_3, can be obtained from the American Smelting and Refining Company† or prepared[12] by reacting indium(III) chloride with ammonium hydroxide, filtering the indium(III) hydroxide, and igniting the hydroxide at 1000°C. in air.

Procedure

An intimate mixture of 2.775 g. (0.01 mole) of indium(III) oxide and 1.718 (0.01 mole) of indium(III) fluoride is ground to a fine powder with an agate mortar and pestle. It is best to carry out this procedure in a dry-bag. The powder is pelletized, and the pellet is placed in a platinum tube. The tube is evacuated and sealed after crimping to exclude air. A torch seal is preferred to a hammer seal because the exclusion of air or moisture during reaction is imperative. The platinum tube is heated to 900°C. for a period of 10 hours. The tube is then cooled to room temperature and opened. A white microcrystalline product is obtained. Larger crystals can be obtained by operating at a slightly higher temperature (1000°C.) and for prolonged heating periods (20–30 hours).

*1870 S. Boulder Ave., Tulsa, Okla. 74119.
†120 Broadway, New York, N.Y. 10005.

The product is air- and moisture-stable. *Anal.* Calcd. for InOF: In, 76.6; F, 12.7. Found: In, 76.8; F, 12.4. The observed density is 6.55 g./cc.,* whereas the calculated theoretical density from the crystallographic data is 6.64 g/cc.

Properties

Indium(III) oxyfluoride is a white crystalline solid which is practically insoluble in all ordinary solvents. It is not hydrolyzed in water at room temperature over a period of 24 hours, and it can be recovered unchanged from boiling water after 1 hour. The pure product was found to decompose at 340°C. in air to yield indium(III) oxide. It is stable to higher temperatures (1100°C.) under an inert atmosphere or in vacuum.

The electrical resistivity data on crystals of indium(III) oxyfluoride indicate a nearly temperature independent conductor (3.6 X 10^{-2} Ω-cm. at room temperature and 1.8 X 10^{-2} Ω-cm. at liquid-helium temperature) with high negative thermoelectric power (-230 μV./°C.). These properties are similar to those observed for some conductive forms of indium(III) oxide.

Single-crystal precession data indicate orthorhombic symmetry with the crystallographic space group *Fddd*. This system is not isostructural with any other known metal oxyfluoride or metal dioxide. The cell dimensions, determined from Guinier data, are $a = 8.370 \pm 1$ A.; $b = 10.182 \pm 1$ A.; and $c = 7.030 \pm 1$ A. The indexed powder data are given in reference 6.

References

1. P. L. Goggin, I. J. McColm, and R. Shore, *J. Chem. Soc. (A)*, **1966**, 1004.
2. H. E. Forsberg, *Acta Chem. Scand.*, 11, 676 (1957).
3. K. Dehnicke, J. Weidlein, and K. Krogmann, *Angew. Chem. Int. Ed. Engl.*, 3, 142 (1964).

*Checkers found 6.60 g./cc. for density.

4. G. Mermant, C. Belinski, and F. Lalau-Keraly, *Compt. Rend. Acad. Sci. Paris,* 263C, 1216 (1966).
5. J. Grannec, J. Portier, R. de Pape, and P. Hagenmuller, *Bull. Soc. Chim. France,* 11, 4281 (1967).
6. B. L. Chamberland, and K. R. Babcock, *Mat. Res. Bull.,* 2, 481 (1967).
7. H. Hahn and W. Nickels, *Z. Anorg. Allgem. Chem.,* 314, 303 (1962).
8. P. Hagenmuller, J. Portier, J. Cadiou, and R. de Pape, *Compt. Rend. Acad. Sci. Paris,* 260C, 4768 (1965).
9. J. Grannec, J. Portier, R. von der Mühll, G. Demazeau, and P. Hagenmuller, *Mat. Res. Bull.,* 5, 185 (1970).
10. A. W. Sleight, J. L. Gillson, and B. L. Chamberland, *ibid.,* 807 (1970).
11. O. Hannebohn and W. Klemm, *Z. Anorg. Allgem. Chem.,* 299, 342 (1936).
12. A. Thiel and H. Luckmann, *ibid.,* 172, 353 (1928).

24. STRONTIUM CHLORIDE PHOSPHATE AND VANADATE(V)

$$a. \quad 10SrCl_2 + 3P_2O_5 + 9HOH \longrightarrow 2Sr_5(PO_4)_3Cl + 18HCl$$
$$b. \quad 4SrCl_2 + V_2O_5 + 3HOH \longrightarrow 2Sr_2(VO_4)Cl + 6HCl$$

Submitted by L. H. BRIXNER*
Checked by R. KERSHAW† and A. WOLD†

The synthesis of the title compositions has been selected as representative of compounds that can readily be prepared by the *flux-reaction technique*. In this technique, a halide melt serves both as a flux and as a constituent component of the basic reaction. The procedure has been described in the literature[1] and has served for the preparation of a variety of ternary oxides, usually in the form of small, well-defined, single crystals. The halide phosphates and vanadates of strontium represent the apatite and spodiosite structures, both interesting compositions from a biochemical and solid-state point of view.

*Central Research Department, E. I. du Pont de Nemours & Company, Wilmington, Del. 19898.
†Brown University, Providence, R.I. 02912.

Procedure

A. PENTASTRONTIUM CHLORIDE TRIS(PHOSPHATE)

The reaction that is fundamental to the flux-reaction technique is the high-temperature hydrolysis of strontium chloride:

$$SrCl_2 + H_2O \longrightarrow SrO + 2HCl$$

At about 1000°C. the ΔH for this reaction is 58 kcal. according to Bichowski.[2] This energy must be supplied by the heat of formation of the compound to be made or else there will be no reaction at all.

For the preparation of $Sr_5(PO_4)_3Cl$ the reaction is given in the above equation *a*. To develop best crystals, the reaction is carried out in two steps: one at 1000°C., and one at 1200°C. Two hundred grams (1.26 moles) of anhydrous strontium chloride (this reagent is obtainable from Inorganic Chemical Company;* if Baker and Adamson† $SrCl_2 \cdot 6H_2O$ is used, it must be dehydrated with dry hydrogen chloride at 150°C. to constant weight) and 5 g. (0.35 mole) of Baker and Adamson reagent-grade phosphorus(V) oxide are mixed and melted in 250-ml. triangle RR recrystallized alumina container.‡ The crucible is placed in a Globar furnace, or any other muffle or box furnace, with a minimum height of heating space of 10 cm. The temperature is raised slowly to 1000°C., and at this temperature a 12-mm.-o.d., 30-cm.-long, recrystallized alumina tube is inserted gradually into this melt so that the final configuration indicated in Fig. 15 is obtained. As indicated in the figure, air that has been saturated with water vapor by passage through a water-filled wash bottle is bubbled through the melt via this tube. The temperature of the wash bottle

---

*Inorganic Chemical Company, 11686 Sheldon St., Sun Valley, Calif. 91352.
†Allied Chemical Company, General Chemical Division, Morristown, N.J. 07960.
‡Morganite Company, 3302 48th Ave., Long Island City, N.Y. 11100.

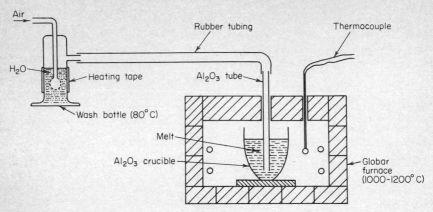

Air →

H₂O

Heating tape

Wash bottle (80°C)

Rubber tubing

Al₂O₃ tube

Melt

Al₂O₃ crucible

Thermocouple

Globar furnace (1000–1200°C)

Fig. 15

is maintained at about 80°C. by wrapping it with heating tape; total flow rate is about 8–9 l./hour. The temperature of the reaction medium is maintained at 1000°C. for one hour and then raised to 1200°C for another hour. Since hydrogen chloride is generated, the reaction should be carried out in a *fume hood*. After 2 hours the crucible is furnace-cooled. When room temperature is attained, the crucible content is water leached in a 2–3-l. beaker, the residue is filtered, washed with water and alcohol, and finally dried at 150°C. About one-half of the product is obtained in the form of highly acicular, finely divided material, whereas the other fraction will be in the form of well-developed single crystalline needles up to 1 cm. long. A typical yield is 15–17 g., which represents 90–95% of theory calculated from equation *a*. *Anal.* Calcd. for $Sr_5(PO_4)_3Cl$: Sr, 57.7; P, 12.25; Cl, 4.6. Found: Sr, 56.7; P, 12.3; Cl, 4.3.

B. DOPED STRONTIUM CHLORIDE PHOSPHATES

1. $Sr_5(P_{0.99}Mn_{0.01}O_4)_3Cl$

This variation of the pure strontium chloroapatite contains a small portion of manganese(V) in the tetrahedral environment and has a brilliant blue color. It can be obtained by simply adding 0.5 g. (0.003 mole) of manganese(II) sulfate monohydrate or manganese(II) carbonate (0.004 mole) to the original mixture of 200 g. (1.26 moles) of strontium chloride and 5 g. (0.35 mole) of phosphorus(V) oxide. *Anal.* Calcd. for $Sr_5(P_{0.99}Mn_{0.01}O_4)_3Cl$: Sr, 57.7; Mn, 0.2; P, 12.1; Cl, 4.7. Found: Sr, 57.0; Mn, 0.2; P, 12.1; Cl, 4.9.

In a typical experiment, 9.8 g. is obtained as coarse needles (58%), whereas 7.1 g. (42%) is in the form of fibers. Total yield is 16.9 g., or 95.2%.

2. $Sr_5(P_{0.99}Cr_{0.01}O_4)_3Cl$

This intensely green-colored chromium(V) doped compound can also be obtained in the same way by adding 0.5 g. (0.003 mole) chromium(III) oxide to the original mixture. *Anal.* Calcd. for $Sr_5(P_{0.99}Cr_{0.01}O_4)_3Cl$: Sr, 57.7; Cr, 0.2; P, 12.2; Cl, 4.7. Found: Sr, 57.1; Cr, 0.2; P, 12.4; Cl, 4.6.

A yellow color of the leach water is normal and results from a small fraction of strontium chromate which also forms.

3. $Sr_{4.75}Eu_{0.25}(PO_4)_3Cl$

This compound constitutes an example of a cation substitution and can be obtained by adding 1.0 g. (0.0029 mole) of europium(III) oxide to the original mixture. It is particularly interesting that under the conditions of the reaction europium(II) is formed. Excitation with a regular ultraviolet lamp of either 2537 or 3660 A. shows intense blue fluorescence.

C. STRONTIUM CHLORIDE VANADATE(V)

By using 5 g. (0.027 mole) of vanadium(V) oxide* together with 200 g. (1.26 moles) of strontium chloride and the same basic procedure as for the chloride phosphate, the reaction product,

*Vanadium Corporation of the United States, Cambridge, Ohio 43725.

strontium chloride vanadate(V), is obtained in the form of thin, transparent platelets rather than needles and structurally is a spodiosite.[3] The equation describing the reaction is given by equation *b*. In a typical experiment, 12 g. of product is obtained, representing 91% of the theoretical yield. *Anal.* Calcd. for Sr_2VO_4Cl: Sr, 53.81; V, 15.64; Cl, 10.89. Found: Sr, 54.0; V, 15.2; Cl, 9.05. This compound also fluoresces in the blue upon ultraviolet excitation 2537 A.

Properties

The $Sr_5(PO_4)_3Cl$ crystals are hexagonal needles with lattice parameters a_h = 9.953 A. and c_h = 7.194 A. The needle axis corresponds to the crystallographic *c* axis. The europium(II) doped sample is a phosphor, readily excitable with electrons, x-rays, and both short and long ultraviolet light. It emits in the blue with a peak at 445 nm. Crystals of strontium chloride vanadate(V) are orthorhombic platelets with lattice constants a = 7.43 A., b = 11.36 A., and c = 6.54 A., with the *b* axis corresponding to the thin dimension of the flakes. Strontium chloride vanadate(V) is a self-activated phosphor giving broadband emission with a peak at 423 nm. when excited with 2537-A. radiation. All compounds are insulators, with resistivities $>10^{12}$ Ω-cm.

References

1. L. H. Brixner, *Mat. Res. Bull.*, 3, 817 (1968).
2. F. R. Bichowski and F. D. Rossini, "The Thermochemistry of the Chemical Substances," p. 126, Reinhold Publishing Corporation, New York, 1936.
3. A. D. MacKay, *Min. Mag.*, 30, 166 (1953).

II. OXIDES

25. NIOBIUM MONOXIDE

$$3Nb + Nb_2O_5 \longrightarrow 5NbO$$

Submitted by T. B. REED* and E. R. POLLARD*†
Checked by L. E. LONNEY,‡ R. E. LOEHMAN,‡ and J. M. HONIG‡

The electric arc has been used since the time of Henri Moissan (1890) in the preparation of refractory inorganic materials. The more recent (1940) development of inert-gas, cold-hearth melting techniques[1] has made arc melting one of the simplest and most versatile methods for the synthesis of those inorganic compounds which are stable up to their melting points. Synthesis by this technique removes volatile impurities and produces dense, homogeneous samples, from which crystals of several millimeters on a side can often be isolated. Crystals thus obtained are suitable for the measurement of chemical and physical properties. The synthesis of NbO has been described below, but the same procedure with minor modifications has been used to prepare other refractory oxides of intermediate valence, such as TiO, Ti_2O_3, Ti_3O_5, VO, and V_2O_3.

Procedure

A laboratory-size arc furnace§ suitable for preparing NbO buttons of up to 30 g. is shown schematically in Fig. 16. The

*Lincoln Laboratory, M.I.T., Lexington, Mass. 02173. Work supported by the U.S. Air Force.

†Present address, Raytheon Corporation, Burlington, Mass.

‡Department of Chemistry, Purdue University, W. Lafayette, Ind. 47907. Work supported by N.S.F. grant, GP 8302.

§Centorr Company, Suncook, N.H. 03275.

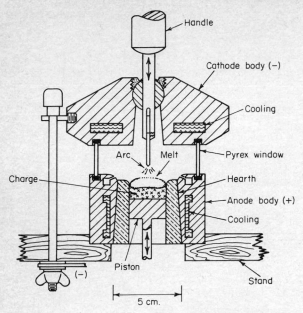

Fig. 16. Schematic diagram of cold-hearth arc-melting furnace with movable piston.

power source is a d.c. welding power supply, capable of supplying
at least 300 amp., which has the drooping-voltage characteristic
used in Heliarc welding. A thoriated tungsten electrode, $\frac{1}{8}$ or $\frac{3}{32}$ in.
in diameter, serves as the cathode, and the water-cooled copper or
graphite hearth as the anode. The charge is placed in the
cylindrical cavity with a movable graphite piston at the bottom.
Water cooling is sufficient to keep the hearth cool while the
high-temperature reaction occurs, and to form a thin protective
layer around the melt, which will protect it from contamination
by the hearth material.

The hearth cavity is filled with high-purity Nb_2O_5 powder and
Nb metal rod* or powder in appropriate proportions. The entire
assembly is flushed with argon gettered over Ti chips or foil at
900°C. to remove any O_2, N_2, or hydrocarbons. A continuous

*Materials Research Corporation, Orangeburg, N.Y. 10962.

argon flow of approximately 2 l/minute is maintained during operation to carry the volatile impurities out of the chamber. The arc is struck by touching the electrode momentarily to the hearth. The electrode is then raised and swiveled until the arc plasma bathes the mixed powders. As the metal melts and reacts with the oxide powder, the material sinks into the crucible and the piston is continuously raised to keep the melt in view. If a simple cup-shaped hearth is employed in place of the piston hearth, the electrode is lowered in the course of the reaction.

After the reaction is complete, the power is raised to melt the entire button and then gradually reduced to ensure slow cooling. After the arc is extinguished, it is desirable to flip the button over with the electrode and then to strike a new arc and remelt the button; this procedure ensures greater homogeneity.

The resulting button exhibits a bright silvery metallic sheen. As long as the oxygen/metal ratio x (in NbO_x) is within the homogeneity range of $0.98 \leqslant x \leqslant 1.02$, single-crystal grains up to 4 mm. on an edge are formed.

The stoichiometry of the sample is determined readily by heating an aliquot portion in air for 16 hours at 800°C. to form Nb_2O_5. The oxygen/metal ratio is calculated from:

$$x = 2.500 - 8.3070 \frac{g_2 - g_1}{g_2}$$

where g_1 and g_2 are the initial and final weights. If $g_1 = 0.5$ g., and g_1 and g_2 are measured to 0.1 mg., then x is known to within ±0.002. Adjustments in x may be made by adding Nb or Nb_2O_5 to the button and remelting.

A typical analysis using the spark-source mass spectrograph shows the following impurities: N = 400, C = 300, Ta = 200, Fe = 90, Mo = 40, V = 10 atomic p.p.m. All other impurities are below the 10-p.p.m. limit; it is interesting that only 1 p.p.m. each of Cu and W have been detected, even though these are the major construction materials of the furnace used.

Properties

The congruently melting composition is $NbO_{1.006}$, which melts at $1940°C$. The lattice parameter of $NbO_{1.000}$ is 4.2111 ± 0.0002 A. at room temperature. The resistivity of NbO_x at $300°C$. and 4.2 K. is approximately 2×10^{-5} and 7×10^{-7} Ω-cm., respectively. The Seebeck coefficient at room temperature varies between $+1.0$ and -1.0 μvolt/degree, depending on x.

References

1. T. B. Reed, *Mat. Res. Bull.*, 2, 349 (1967).
2. E. R. Pollard, Electronic Properties of Niobium Monoxide, Ph.D. thesis, Dept. of Material Sciences, M.I.T. (1968).

26. MAGNESIUM CHROMITE

(Magnesium Chromium(III) Oxide)

$$Cr_2O_3 + WO_3 \longrightarrow Cr_2WO_6$$
$$MgF_2(l) + Cr_2WO_6(l) \longrightarrow MgCr_2O_4(c) + WO_2F_2(g) \uparrow$$

Submitted by W. KUNNMANN*
Checked by D. BELLAVANCE†

Magnesium chromium(III) oxide can be synthesized by direct solid-state reaction of the component oxides Cr_2O_3 and MgO at an elevated temperature. The product is generally of poor quality owing to the slow rates of diffusion of the cations that lead to inhomogeneities on an atomic scale. Whipple and Wold[1] have shown that a superior product is obtained by the controlled reduction of magnesium dichromate, $MgCr_2O_7$, where the magnesium and chromium ions are premixed on an atomic scale. Low-temperature neutron-diffraction results[2] show a significant

*Brookhaven National Laboratories, Upton, N.Y. 11973.
†Materials Research Center, Allied Chemical Corporation, P.O. Box 3004, Morristown, N.J. 07960.

difference between samples prepared by the two methods. It was observed that the product obtained by the dichromate-reduction method possessed both a high- and low-temperature magnetic structure, whereas samples prepared by solid-state reaction always yielded only the low-temperature magnetic structure. Magnesium chromium(III) oxide prepared by the metathesis reaction to be described shows magnetic structure transformations identical with those observed for samples prepared by the dichromate-reduction method.[2]

A. DICHROMIUM TUNGSTEN(VI) OXIDE, Cr_2WO_6

As originally synthesized by Bayer,[3] dichromium tungsten(VI) oxide is prepared easily by direct solid-state reaction of the component oxides, chromium(III) oxide and tungsten(VI) oxide.

Procedure

One-tenth formula weight of reagent-grade chromium(III) oxide (15.202 g.) is dried overnight at $100°C$. or higher, and 23.186 g. (0.1 mole) of reagent-grade tungsten(VI) oxide is prefired at $500°C$. for 4 hours. The reactants are mixed intimately by being ground together in an agate mortar with an agate pestle. The mixture so obtained is pressed into pellets, with a hand press, and placed in a platinum crucible equipped with a cover. The crucible is placed in a muffle type of furnace. The furnace temperature is then raised to $1000°C$., and reaction is permitted to proceed overnight (12–14 hours). The furnace is subsequently cooled to room temperature, and the partially reacted pellets are crushed. The procedures of grinding, pressing into pellets, and firing are repeated. After the second firing, the material has undergone sufficient reaction to be used for the metathetical reaction to magnesium chromite. For a high-quality preparation of dichromium tungsten(VI) oxide, however, the recycling processes of grinding, pressing, and firing must be repeated several more times before high-sensitivity x-ray diffraction techniques indicate a single phase is present. If a high-quality product is desired, care must be taken to secure a tight-fitting lid on the platinum crucible since some tungsten(VI) oxide will volatilize at high temperature.

Properties

Dichromium tungsten(VI) oxide is a deep-purple powder possessing the trirutile type of structure. The compound crystallizes in the space group $P 4_2/mnm$. with $Z = 2$, $a_0 = 4.582$ A., $c_0 = 8.870$ A.

B. MAGNESIUM CHROMIUM(III) OXIDE, $MgCr_2O_4$
(Magnesium Chromite)

Although an equimolar mixture of molten magnesium fluoride and dichromium tungsten(VI) oxide react at 1270°C. (in an air atmosphere) to form magnesium chromium(III) oxide crystals, the metathetical reaction is best carried out by using an excess of magnesium fluoride and increasing the reaction temperature to 1400°C. The best results are obtained at a ratio of magnesium fluoride to dichromium tungsten(VI) oxide of 2:1 or 3:1.

Procedure

Typically, 38 g. (0.1 mole) of dichromium tungsten(VI) oxide and 18.7 g. (0.3 mole) of oven-dried, reagent-grade, magnesium fluoride are mixed and placed in a platinum crucible fitted with a cover.

Since the metathetical reaction between the fluoride and tungstate will also proceed in the solid state (although at a much slower rate), it is desirable to have the reaction mixture attain the liquid state in as short a time as possible. If a furnace capable of attaining 1300–1400°C. in a few hours is not available, the following procedure may be employed. The furnace (such as the Harper* furnace model, HOU 6610 M 30) is preheated to 1300°C., the furnace chamber opened, the sample rapidly placed inside, the furnace chamber closed, and the temperature raised to 1400°C.

---

*Any suitable box furnace capable of attaining a temperature of 1400°C. may be used.

■ *Caution. Extreme care should be exercised when operating in front of the heating chamber since the radiation is of such intensity as to cause burns on the exposed hands and face and to char clothing. Very long tongs (3–4 ft.) should be used in addition to protecting the hands with asbestos gloves, wearing a protective face mask, and using a radiation shield such as a makeshift bib of aluminum foil—shiny side facing furnace—to protect clothing.*

The temperature is held constant at 1400°C., and the metathesis of the tungstate to the oxyfluoride proceeds slowly to bring about the gradual production of magnesium chromium(III) oxide, which subsequently is precipitated. After approximately 24 hours, the reaction will be about 40% complete. As pointed out previously, however, physical considerations have a strong influence, and the largest crystals will be approaching millimeter size. Generally, about 3 days at 1400°C. are required to produce a measurable yield of millimeter-sized crystals. By scaling the reacting mixtures up by a factor of 10 and carrying out the reaction at a temperature of 1500°C. for 14 days, magnesium chromium(III) oxide octahedra approaching centimeter size can be obtained.

■ *Note. Gaseous tungsten oxyfluoride, WO_2F_2 (difluoro-dioxotungsten), which is produced during reaction, is a very corrosive chemical at high temperature. It is recommended, therefore, that an old alumina muffle be used or that adequate protection against attack be provided. This may be accomplished by inserting the platinum crucible containing the reactants into a heavily walled alumina crucible, which will itself react with the tungsten oxyfluoride, or by providing the muffle with protective inserts. These are commercially available by the furnace manufacturing companies primarily for use with silicon carbide muffles.*

After the reaction has proceeded for the desired interval of time, the platinum crucible containing the reacted material is rapidly cooled to room temperature. This is most easily accomplished by shutting off the power to the furnace and quickly removing the platinum crucible when the temperature has dropped to approximately 900°C. (Removal of the crucible may be made at higher temperatures if desired; the crystals of $MgCr_2O_4$ will withstand the thermal shock quite well.) The reason for rapid cooling is to minimize the reaction of atmosphere oxygen with the magnesium chromium(III) oxide to form the chromate. If a 5–10% loss of the product is not objectionable, the crucible may be cooled in the furnace.

Removal of the product from the unreacted magnesium fluoride and chromium tungstate, Cr_2WO_6, may be accomplished by leaching with hot concentrated hydrochloric acid in a fume hood. This will slowly etch the product free, and overnight treatment will be necessary for complete removal. It will be found that no noticeable attack of the chromite crystals will occur during the leaching process. However, the Cr_2WO_6 is decomposed to tungstic acid (insoluble) and soluble chromium(III) chloride, while most of the magnesium fluoride is dissolved. A small portion of the magnesium fluoride reacts with the tungstic acid to form a small amount of magnesium tungstate. The liquid is decanted and discarded, and the solid residue is washed with distilled water, which is also decanted and discarded. A hot saturated solution of sodium hydroxide is then added to the solid residue. The mixture is stirred for approximately one hour, and the liquid subsequently decanted and discarded. The solid product is then washed with distilled water. Alternate treatment with hot acid leaching and hot base washing ultimatedly yields a clean product of magnesium chromite crystals.

A rapid, alternative method of crystal separation is the addition of solid potassium chloride (an amount necessary to fill the crucible) directly into the untreated, fused mass (obtained after cooling to room temperature), followed by heating the new mixture to its fusion point by means of a Meker burner in a fume hood and maintaining this temperature for 1 or 2 hours. The molten potassium chloride will rapidly dissolve the magnesium fluoride and decompose the tungstate. The crucible is allowed to cool, and the fused mass is treated as described in the previous section. However, dissolving of the flux will now be much more rapid.

Chemical analysis of the magnesium chromite crystals is difficult because of the insolubility of the sample in most reagents. Long-term boiling in concentrated mineral or oxidizing acids has little effect. Disulfate (pyrosulfate) or sodium hydroxide fusion for several hours is equally ineffectual. However, if the crystals are ground to a fine powder (<250 mesh), they can then be brought into solution by alternating disulfate and sodium hydroxide fusions. The chromium-to-magnesium ratio is very close to 2:1 (2.01:1). Diffraction data with refined site occupancy support this conclusion.[2]

Properties

Magnesium chromium(III) oxide crystals are deep-green octahedra possessing the normal, cubic, spinel structure, a_0 = 8.335 A., and having a melting point of approximately 2390°C.

References

1. H. Whipple and A. Wold, *J. Inorg. Nucl. Chem.*, 24, 23 (1962).
2. H. Shaked, J. M. Hastings, and L. M. Corliss, *Phys. Rev. B*, 1 (7), 3116 (1970).
3. G. Bayer, *J. Am. Ceram. Soc.*, 43, 495 (1960).

27. SILVER FERRATE(III)

$$Ag_2O + Fe_2O_3 \xrightarrow[\text{40,000 p.s.i. aq. NaOH}]{400°C} 2AgFeO_2$$

Submitted by WILLIAM J. CROFT* and NIGEL C. TOMBS†
Checked by R. D. SHANNON‡

The formation of silver ferrate(III) of empirical formula $AgFeO_2$ by reactions with iron hydroxides and silver compounds in

*Army Materials and Mechanics Research Center, Watertown, Mass. 02172.
†NASA Research Center, Moffett Field, San Jose, Calif. 95100.
‡McMaster University, Hamilton, Ontario, Canada; on leave of absence from Central Research Department, E. I. du Pont de Nemours & Company, Wilmington, Del. 19898.

aqueous suspension has been reported previously.[1] The initial interest of the authors was in the use of the red-colored silver compound in identification of specific iron hydroxide compounds. The products obtained were apparently poorly crystallized or inhomogeneous or both, and the reported x-ray powder diffraction data consisted of only nine reflections. The unit cell deduced from these results was rhombohedral, a_R = 4.61 A., α = 85°25'. The method described here is the reaction of a mixture of silver oxide and α-iron(III) oxide under hydrothermal conditions in basic solutions. The silver ferrate(III) produced is in the form of single crystals from which optical and improved x-ray data were obtained.

Procedure

The reaction system consists of a mixture of silver(I) oxide and α-iron(III) oxide in 3 M aqueous sodium hydroxide solution. A slight excess of silver oxide over the stoichiometric requirement ensures that no unreacted iron oxide will remain. The reaction system is sealed in a platinum tube of 6-mm. i.d., 0.1-mm. wall thickness, and 5-cm. length. The sealing is made by crimping the tube ends and welding the seams with a carbon electrode arc. The sealed tube is heated in a Stellite alloy, pressure vessel of conventional design. The reaction is complete essentially in 72 hours at 400°C. and 40,000 p.s.i. The small deliberate excess of silver oxide can be removed readily by treatment with dilute nitric acid. The reaction product consists of shiny platelets up to about 2 mm. on a side and about 0.2 mm. thick, together with finer crystallized material. Both the platelets and the smaller crystallites give the same x-ray powder pattern.

Analysis and Properties

Emission spectrographic analysis shows only silver and iron as major constituents. Of special interest is the fact that sodium is present to the extent of less than 1 p.p.m., although the crystallization takes place in sodium hydroxide solution. Up to 0.3% sodium has been reported in preparations from co-precipitated hydroxides.[2] Silver and iron can be analyzed by flame photometry. The sample and standard are dissolved in concentrated sulfuric acid, which is then diluted. The analytical results confirm a silver-to-iron ratio of unity.

When heated in air, $AgFeO_2$ crystals are stable up to about 700°C., at which temperature they decompose with loss of oxygen to yield metallic silver and α-iron(III) oxide.

Crystallographic Data

Many of the thinner platelets are transparent and a rich ruby-red color in transmission. They give a uniaxial negative interference figure. The crystals are too deeply colored for refractive index determination, but the mean value is greater than 2.

The density can be measured in toluene, with a modified Berman balance. A value of 6.39 g./cc. is obtained. The calculated value of 6.56 is obtained by using a value of 3 units of $AgFeO_2$ per hexagonal unit cell.

Recently, a detailed study[4,5] of ABO_2 compounds confirms that $AgFeO_2$ has the $CuFeO_2$ (delafossite) structure. The cell dimensions, determined with a Guinier camera, are found to be $a =$ 3.0391 ± 2 A. and $c =$ 18.590 ± 2 A. Single crystals are used to measure resistivity as a function of temperature.[6] Silver ferrate(III) exhibits semiconductor behavior with an activation energy of 0.7 e.V. An unusual anisotropy in resistivity is found with $\rho_{(\perp c)} = 3 \times 10^7$ Ω-cm. and $\rho_{(\| c)} = 2 \times 10^{10}$ Ω-cm.

References

1. A. Krause, Z. Ernst, St. Gawryck, and W. Kocay, *Z. Anorg. Allgem. Chem.*, **228**, 352 (1936).
2. A. Krause and K. Pilawski, *ibid.*, **197**, 301 (1931).
3. International Tables for X-ray Crystallography (1952).

4. R. D. Shannon, D. B. Rogers, and C. T. Prewitt, *Inorg. Chem.*, **10**, 713 (1971).
5. C. T. Prewitt, R. D. Shannon, and D. B. Rogers, *ibid.*, 719 (1971).
6. D. B. Rogers, R. D. Shannon, C. T. Prewitt, and J. L. Gillson, *ibid.*, 723 (1961).

28. BARIUM TITANATE, $BaTiO_3$

(Barium Titanium(IV) Oxide)

$$BaCO_3 + TiO_2 \longrightarrow BaTiO_3 + CO_2$$

Submitted by FRANCIS S. GALASSO*
Checked by MICHAEL KESTIGAN†

Barium titanate, $BaTiO_3$, is probably the most widely studied ferroelectric oxide. Extensive studies were conducted on this compound during World War II in the United States, England, Russia, and Japan, but the results were not revealed until after the war. Barium titanium(IV) oxide was found to be a ferroelectric up to a temperature of 120°C., which is its Curie point. Above 120°C., barium titanium(IV) oxide has the cubic perovskite structure, and below this temperature the oxygen and titanium ions are shifted and result in a tetragonal structure with the *c* axis approximately 1% longer than the *a* axis. Below 0°C., the symmetry of barium titanate becomes orthorhombic, and below −90°C., it becomes trigonal.

Because barium titanate has interesting properties, many methods have been used to grow single crystals of this compound. One of the most popular techniques, using a potassium fluoride flux, was first employed by Remeika.[1]

Procedure

Crystals of barium titanate can be grown with a mixture of 30% barium titanate by weight and anhydrous potassium fluoride. In a

*United Aircraft Research Laboratories, East Hartford, Conn. 06108.
†Sperry Rand Research Center, Sudbury, Mass. 01776.

specific procedure, 28 g. of barium titanate [or 23.6 g. (0.1 mole) of barium carbonate and 9.6 g. (0.12 mole) of titanium(IV) oxide] is placed in a 50-ml., platinum crucible and covered with 66 g. of anhydrous potassium fluoride. The cover is placed on the crucible, and the crucible and contents are heated to 1160°C. When barium carbonate and titanium(IV) oxide are used with potassium fluoride, the contents of the crucible are first slowly melted before the cover is placed on tightly and the mixture heated to 1160°C. After being held for 12 hours at 1160°C., the crucible is cooled 25°/hour to 900°C., and the flux is poured off. The crystals then are annealed by slowly cooling them to room temperature, and they are removed by soaking the crucible contents in hot water.

Properties

Crystals grown by this technique are in a form called *butterfly* twins. The twins consist of two isosceles right triangles with a common hypotenuse.[2] Normally the composition plane is (111), the crystal faces are ⟨100⟩ and the angle between the wings is approximately 38°.

The dielectric constant of barium titanate, along [001] is about 200 and along [100] it is 4000 at room temperature.[3] The spontaneous polarization at room temperature is 26 × 10^{-6} C./cm.^{2}, and the value of the coercive field has been found to vary from 500 to 2000 volts/cm. The crystal structure of barium titanate at room temperature can be represented by a tetragonal unit cell with size of a_0 = 3.992 A., and c_0 = 4.036 A., but the symmetry becomes cubic above 120°C., at which temperature the crystals no longer exhibit ferroelectric properties.

References

1. J. P. Remeika, *J. Am. Chem. Soc.*, **76**, 940 (1954).
2. R. C. DeVries, *J. Amer. Cer. Soc.*, **42**, 547 (1959).
3. W. J. Merz, *Phys. Rev.*, **76**, 1221 (1949).

29. BISMUTH TITANATE, $Bi_4Ti_3O_{12}$
(Bismuth Titanium Oxide)

$$2Bi_2O_3 + 3TiO_2 \longrightarrow Bi_4Ti_3O_{12}$$

Submitted by FRANCIS S. GALASSO*
Checked by MICHAEL KESTIGIAN†

Bismuth titanate, $Bi_4Ti_3O_{12}$, is one of a class of ferroelectrics with the general formula $(Bi_2O_2)^{2+}(A_{x-1}B_xO_{3x+1})^{2-}$, where A is a monovalent or divalent element, B is Ti^{4+}, Nb^{5+}, or Ta^{5+}, and x can have values of 2, 3, 4, etc.[1-3] The crystal structure of these compounds consists of Bi_2O_2 layers and stacks of perovskite-like layers perpendicular to the [001] axis. There are three "perovskite" layers between two bismuth oxygen layers in the structure of $Bi_4Ti_3O_{12}$, two perovskite layers on $ABi_2B_2O_9$ compounds, and four in $ABi_4Ti_4O_{15}$ compounds. Since the perovskite layers consist of TiO_6, TaO_6, or NbO_6 octahedra, spontaneous polarization can take place in the plane of these layers.

Procedure

To form a $\frac{3}{8}$-in.-diam. pellet of $Bi_4Ti_3O_{12}$, a sample is first made by mixing 0.47 g. (1 mmole) of bismuth oxide and 0.12 g. (1.5 mmole) of titanium(IV) oxide. The mixture is pressed into pellets with a calorimeter pellet press. The pellets are placed in Alundum or zircon boats and heated at 700°C. for 4 hours. Then the pellets are ground, pressed again, and reheated at a temperature of 920°C. for one hour.

*United Aircraft Research Laboratories, East Hartford, Conn. 06108.
†Sperry Rand Research Center, Sudbury, Mass. 01776.

These pellets can be ground for powder x-ray studies, or electrodes can be evaporated on the flat surfaces of the pellets to produce samples for ferroelectric measurements. However, for more meaningful measurements single crystals of bismuth titanate are required.

Single crystals of $Bi_4Ti_3O_{12}$ can be grown by the following technique.[4] A charge of 210 g. (0.45 mole) of bismuth oxide and 11.0 g. (0.14 mole) of titanium(IV) oxide in a 50-ml. platinum crucible is heated at 1200°C. for one hour. The charge is placed in the crucible by filling the crucible, melting, adding more charge, and remelting. The furnace is slowly cooled at a rate of 40°/hour. The crystals can be removed by dissolving the bismuth oxide in hydrochloric acid.

If a shallow platinum container ($6 \times 4 \times \frac{1}{2}$ in.) is used instead of a typical platinum crucible, larger, higher-optical-quality, single crystals are obtained.

Properties

The crystals grown in this manner are in the form of clear sheets. The symmetry of the crystals is pseudotetragonal with a cell size of $a_0 = 3.841$ A. and $c_0 = 32.83$ A. Electrodes can be evaporated, or indium amalgam can be applied to the flat surfaces of the crystals, to produce samples for measurements. The d.c. resistance of the crystals is about 10^{12} Ω-cm. They exhibit ferroelectric hysteresis loops up to the Curie temperature of 643°C.

References

1. B. Aurivillius, *Arkiv. Kemi,* **1,** 463 (1949).
2. B. Aurivillius, *ibid.,* 499 (1949).
3. B. Aurivillius, *ibid.,* 519 (1959).
4. L. G. Van Uitert and L. Egerton, *J. Appl. Phys.,* **32,** 959 (1961).

30. CADMIUM RHENIUM(V) OXIDE, $Cd_2 Re_2 O_7$ *

$$2Cd + Re_2 O_7 \longrightarrow Cd_2 Re_2 O_7$$

Submitted by JOHN M. LONGO,† PAUL C. DONOHUE,‡
and DONALD A. BATSON §
Checked by D. P. KELLY‡ and A. W. SLEIGHT‡

Cadmium rhenium(V) oxide has been prepared by two methods.[1] The first involves the heating of finely ground powders according to the equation $\frac{5}{3}ReO_3 + \frac{1}{3}Re + 2CdO \rightarrow Cd_2 Re_2 O_7$. This reaction is carried out in a sealed silica ampul at 1000°C. for 24 hours. Some small single crystals are formed on the walls of the ampul. The second method gives more and larger crystals,[1] and an adaptation of this method is described below. Cadmium rhenium(V) oxide is an interesting compound since it is one of the few stoichiometric mixed metal oxides with metallic conductivity.

Procedure

This preparation of cadmium rhenium(V) oxide consists of two steps. The first step is the oxidation of rhenium metal in a stream of oxygen to its heptaoxide $Re_2 O_7$. The second step is the vapor-phase reaction of cadmium metal and rhenium(VII) oxide.

Both steps are carried out in a silica tube (1-cm. i.d. $\times$ 60 cm. long). A thin wad of glass wool is inserted about 20 cm. from the oxygen input end of the tube. Powdered rhenium metal, 0.4655 g.

*The Lincoln Laboratory portion of this work was sponsored by the U.S. Air Force.
†Esso Research and Engineering, Box 45, Linden, N.J. 07036.
‡E. I. du Pont de Nemours & Company, Experimental Station, Wilmington, Del. 19898.
§M.I.T., Lincoln Laboratory, Lexington, Mass. 02173.

(0.0025 mole), is placed in the tube on the upstream side of the glass wool, which acts both as a stop for the rhenium metal and a filter for the rhenium(VII) oxide produced. The cadmium metal, in the form of flakes, is cleaned with 3 N HCl, washed in water, and dried with acetone. Two or three pieces of the metal are cut to weight, 0.2810 g. (0.0025 mole), and placed about 8 cm. from the exit end of the tube. Both ends of the tube are connected with Tygon tubing to Drierite water traps. Oxygen entering through a two-way stopcock is allowed to flow through the tube for a few minutes, after which the tube is placed in a clamshell furnace 12 in. long. The rhenium powder is at the center of the hot zone, which is maintained at 350°C., and the end of the tube containing the cadmium is outside the furnace. The rhenium(VII) oxide vapor produced by the oxidation of the rhenium powder condenses to form a ring at the end of the furnace. The oxygen flow rate is important and should be about 0.08 c.f.h. for the size of tube prescribed here. A faster rate carries the rhenium(VII) oxide too far downstream before condensation, and a slower rate oxidizes the rhenium metal too slowly. When all the rhenium has been oxidized, the tube is repositioned in the furnace (with the oxygen still flowing) so that the rhenium(VII) oxide is resublimed to a region about 8 cm. upstream from the cadmium metal. The final rhenium(VII) oxide should be bright canary yellow and crystalline.

The Tygon tubing at the exit end is clamped off, and the two-way stopcock at the entrance is turned to a vacuum line. After the whole system has been evacuated, the silica tube is sealed first at the exit, and secondly beyond the rhenium(VII) oxide to give a 12–15-cm. ampul. Care must be taken not to melt the cadmium nor to resublime the rhenium(VII) oxide. The sealing operation is done with the tube in a horizontal position to prevent movement of the material.

The ampul is fired in a furnace at 800°C. for about 20 hours. The crystals formed vary in size (1–4 mm. in largest dimension), number, and position, depending on the temperature profile of the furnace and the relative position of the two reactants. Because the crystals tend to adhere to the inside of the silica tube, only about 80% of the product can be recovered conveniently. If moisture is present in the ampul, the reaction is incomplete, and rhenium(IV) oxide and cadmium oxide are found as impurities. *Anal.* Calcd. for $Cd_2Re_2O_7$: Re, 52.5; Cd, 31.7; O, 15.8. Found: Re, 51.5; Cd, 32.0; O, 16.8.

Properties

Cadmium rhenium(V) oxide is deep violet in color. Its crystalline habit is roughly octahedral with the triangular faces (111) having their corners cut. It is stable in air and attacked slowly in strongly oxidizing media. The crystals are treated with dilute nitric acid to ensure a pure product since the impurities are soluble in this acid.

The crystal structure of cadmium rhenium(V) oxide, as determined by single-crystal technique,[1] is of the face-centered cubic pyrochlore type ($a = 10.219$ A.). The only positional parameter for the 48 (f) oxygens is $x = 0.309 \pm 0.007$ when rhenium is at the origin. The density, determined pycnometrically, is 8.82 ± 0.03 g./cc., compared with the theoretical value of 8.83 g./cc. for $Z = 8$. The resistivity between 4.2 K and room temperature is very low (10^{-3}–10^{-4} Ω-cm.) and has a positive temperature coefficient. Over the same temperature range the magnetic susceptibility is low and temperature-independent. These properties indicate that cadmium rhenium(V) oxide exhibits metallic conductivity.

Reference

1. P. C. Donohue, J. M. Longo, R. D. Rosenstein, and L. Katz, *Inorg. Chem.*, **4**, 1152 (1965).

31. MOLYBDENUM(IV) OXIDE AND TUNGSTEN(IV) OXIDE SINGLE CRYSTALS

$$2MO_3 + M \longrightarrow 3MO_2$$

$$MO_2 + I_2 \underset{T_1}{\overset{T_2}{\rightleftharpoons}} MO_2 I_2$$

$$M = Mo, W$$

Submitted by LAWRENCE E. CONROY* and LINA BEN-DOR†
Checked by R. KERSHAW‡ and A. WOLD‡

The dioxides of molybdenum and tungsten crystallize in distorted forms of the rutile structure. The usual method of preparation is the reduction of the trioxide with the metal at 900–1000°C. *in vacuo* or under an inert atmosphere. This procedure yields only microcrystalline products. Extended periods of reaction are necessary to avoid contamination with oxides intermediate between the trioxide and dioxide. Hägg and Magneli[1,2] have shown that a series of such oxides can be identified in the products of partial reduction of the trioxides. Single crystals of molybdenum-(IV) oxide may be produced by electrolytic reduction of Na_2MoO_4–MoO_3 melts.[3] The synthesis described below makes use of the chemical-transport technique[4] to convert the microcrystalline dioxide to single crystals sufficiently large for electrical and crystallographic investigations. The principles of this method are described in earlier volumes of this series.[5-7]

Procedure

The pure molybdenum(VI) or tungsten(VI) oxide is dried by heating in air at 500°C. for one hour. Approximately 10–15 g. of

*University of Minnesota, Minneapolis, Minn. 55455.
†The Hebrew University, Jerusalem, Israel.
‡Brown University, Providence, R.I. 02912.

the 2:1 molar mixture of the trioxide and corresponding metal is mixed thoroughly and ground together in a mortar. The mixture is transferred to an alumina or silica combustion boat and heated under purified argon gas. Alternatively, the sample may be sealed *in vacuo* in a silica ampul. The $2MoO_3$–Mo mixture is converted to molybdenum(IV) oxide in 70 hours at 800°C. The $2WO_3$–W mixture is converted to tungsten(IV) oxide in 40 hours,* at 900°C. Molybdenum(IV) oxide is produced as a brown-violet powder, and tungsten(IV) oxide as a bronze powder.

■ *Caution. The danger of implosion or explosion is always present when sealed glass vessels are heated, however carefully. Use protective eye and face covering or a protective shield for the apparatus.*

Silica or Vycor ampuls are satisfactory for chemical transport. Optimum dimensions are 15 cm. long and 2.5 cm. o.d., holding a sealed volume of approximately 75 ml. Of the metal dioxide, 4 g. is added to the ampul through a long-stem funnel, along with the required quantity (see Table I) of purified iodine. Several techniques for addition of iodine to a transport ampul are described in Volume XII (p. 161) of *Inorganic Syntheses*. The ampul is then sealed.† To promote the growth of large crystals, it is desirable to transfer the charge of dioxide to the seal-off end (hot zone) of the ampul and to allow crystals to grow at the hemispherical end (growth zone). The ampul is then placed in a cold two-zone transport furnace.‡ The growth zone is heated to 900°C. for at least 12 hours, while maintaining the charge zone at 500–600°C.

This procedure minimizes seed sites to promote the growth of a small number of large crystals. The temperature of the charge zone

---

*These extended heating periods are necessary for complete conversion to the dioxides. During the early stages of heating, the primary products are the phases Mo_4O_{11} and $W_{18}O_{49}$,[1] which transport more readily than MoO_2 and WO_2 and thus interfere with the growth of good single crystals of the last two compounds.

†Vycor tubing that is constricted to <10-mm. o.d. may be sealed with an oxygen-gas flame. Silica tubing or larger-diameter Vycor tubing may require an oxy-hydrogen flame.

‡Such furnaces are described in *Inorganic Syntheses*, **11**, 5 (1968); **12**, 161 (1970).

TABLE 1   Optimum Growth Conditions for MoO$_2$ and WO$_2$

	MoO$_2$	WO$_2$
Ampul dimensions	15 cm. long × 2.5 cm. o.d.	15 cm. long × 2.5 cm. o.d.
Ampul volume	*ca.* 75 ml.	*ca.* 75 ml.
Quantity of MO$_2$ powder	4 g.	4 g.
Iodine concentration	4 mg./ml.	1 mg./ml.
Transport temperatures	900 → 700°C.	900 → 800°C.
Transport time	6–7 days	3 days
Cooling rate, growth zone	10°C./hour	10°C./hour
Temperature programming	Necessary	Not necessary

is then increased to the appropriate temperature listed in Table I, and the growth zone is cooled at the rate of 10°C./hour for the stipulated time period. For the growth of larger crystals of molybdenum(IV) oxide, it is essential that the cooling be very uniform. Therefore, some type of programmed cooling procedure is necessary. Many commercial temperature controllers permit such programmed procedures. The growth of tungsten(IV) oxide crystals is much less sensitive than that of molybdenum(IV) oxide crystals to the rate of cooling. Under the transport conditions specified in Table I, the transport of the dioxide powder is essentially quantitative. The dimensions of the largest crystal obtained by this procedure are roughly 1 × 3 × 3 mm.[3]

Properties

The molybdenum(IV) oxide crystals produced by this method are thick, oblong needles or thick platelets having a brown-violet metallic luster. The cell edges are close to those reported for the polycrystalline powders,[7] monoclinic crystals with a = 5.60 A., b = 4.86 A., c = 5.63 A., β = 120.95°. The crystals are excellent metallic conductors.

The tungsten(IV) oxide crystals obtained by this procedure are polyhedra having bronze metallic luster. The crystallographic data agree well with those reported for the polycrystalline compounds;[8] monoclinic crystals with a = 5.56 A., b = 4.89 A., c = 5.66 A., β = 120.5°. Tungsten(IV) oxide is a good metallic conductor.

Chemical analyses of the oxygen content of these crystals, carried out by (1) careful oxidation to molybdenum(VI) oxide in a stream of oxygen and (2) carbon reduction of the metal, yielded the following data. *Anal.* Calcd. for MoO_2 : O, 25.01. Found, 25.1. Mo:O ratio, 1:2.01. Calc. for WO_2 : O, 14.83. Found, 14.9. W:O ratio, 1:2.01.

References

1. G. Hägg and A. Magneli, *Arkiv. Kemi,* 19A(2), 1(1944).
2. A. Magneli, *Nova Acta Regiae Soc. Sci. Upsal.,* 14(8), 13(1949).
3. A. Wold, W. Kunnmann, R. J. Arnott, and A. Ferretti, *Inorg. Chem.,* 3, 545 (1964).
4. H. Schäfer, "Chemical Transport Reactions," Academic Press, Inc., New York, 1964.
5. A. G. Karipedes and A. V. Cafiero, *Inorganic Syntheses,* 11, 5 (1968).
6. L. E. Conroy, *ibid.,* 12, 158 (1970).
7. D. B. Rogers, S. R. Butler, and R. D. Shannon, *ibid.,* 13, 135(1972).
8. A. Magneli, G. Andersson, B. Blomberg, and L. Kihlborg, *Anal. Chem.,* 24, 1998 (1952)

III. CHALCOGENIDES

32. RARE-EARTH SESQUISULFIDES, Ln_2S_3

$$2Ln + 3S \longrightarrow Ln_2S_3$$
$$3Ln + 3S + 3I \longrightarrow 3LnSI \longrightarrow Ln_2S_3 + LnI_3$$

Submitted by A. W. SLEIGHT* and D. P. KELLY*
Checked by R. KERSHAW† and A. WOLD†

Rare-earth sesquisulfides have generally been prepared by the reaction of hydrogen sulfide with rare-earth oxides. However, such

*Central Research Department, E. I. du Pont de Nemours & Company, Wilmington, Del. 19898.
†Brown University, Providence, R.I. 02912.

a procedure frequently gives oxysulfides or nonstoichiometric sulfides. Direct reaction of the elements readily gives pure stoichiometric sesquisulfides, and this procedure is easily modified to give crystals.[1]

Procedure

High-purity sulfur, iodine, and rare-earth metals are commercially available. The sponge or powder form of the rare-earth metal is easiest to work with, but such forms are not satisfactory (and usually not available) for the more active rare-earth metals. Thus, although the examples below assume that Gd sponge or powder is used, ingot forms are used for La, Pr, Nd, Tb, Tm, Yb, and Lu. Such ingots must be freshly cut or filed.

Polycrystalline gadolinium sesquisulfide can be prepared from 1.5725 g. (0.01 mole) of gadolinium and 0.4810 g. (0.015 mole) of sulfur. The reactants are not ground together or in any way intimately mixed. They are simply placed together in a 10-mm. silica tube which has been sealed at one end. The tube is evacuated and sealed to produce an ampul which is about 20 cm. long. The ampul is placed in a two-zone or gradient furnace with all reactants initially in the hot end. The hot-zone temperature is raised to 400°C., while the other zone is maintained at about 100°C. The sulfur quickly moves to the cooler zone, and thereafter vapors of sulfur will react with the gadolinium in a controlled fashion. (■ *Caution. If the entire ampul is heated to 400°C., a violent reaction will occur, with attack on the ampul, which will probably break.*) When all the sulfur is consumed, the ampul is transferred to a muffle furnace and heated at 1000°C. for about 10 hours.

Single crystals of gadolinium sesquisulfide may be prepared from 1.5725 g. (0.01 mole) of gadolinium, 0.3206 g. (0.01 mole) of sulfur, and 1.2690 g. (0.01 mole) of iodine. The reactants are sealed in an evacuated silica tube as described above. The vacuum should be applied no longer than necessary before sealing because of the volatility of iodine. The ampul is then heated in a two-zone or gradient furnace exactly as described above. When all the sulfur and iodine are consumed, the ampul is removed from the furnace. At this point the sample will be basically GdSI. However, this compound will decompose when heated at higher temperatures. Crystals of gadolinium sesquisulfide will grow in a gadolinium iodide melt if the ampul is held at 1100–1200°C. for 20 hours or longer. The crystals may be washed free of gadolinium iodide by using alcohol or water-alcohol mixtures. Red rods several millimeters in length are obtained.

Polycrystalline rare-earth sesquisulfides have been prepared[1] by this method for La, Ce, Pr, Nd, Sm, Gd, Tb, Dy, Ho, Er, Tm, Yb, Lu, and Y. Europium sesquisulfide does not exist. For the more reactive rare-earth metals (La to Sm), the silica ampul will be severely attacked unless protected. This may also be a problem for other rare earths if high temperatures and long heating times are employed. Carbon is the most suitable material for protecting the silica in these syntheses. A graphite crucible may be used, but it is generally satisfactory simply to coat the inside of the silica tube with carbon by the pyrolysis of benzene. Benzene is poured into the silica tube, which is closed at one end; it is then poured back out with the residue left clinging to the tube. The tube is placed in a furnace at 800°C. for a few minutes.

Crystals of rare-earth sesquisulfides have been prepared by this method except when Ln is La, Er, Tm, or Y. In these cases the LnSI compounds are stable even at 1250°C. *Anal.* Calcd. for Gd_2S_3: Gd, 76.58; S, 23.42. Found: Gd, 76.4; S, 22.2.

Properties

The rare-earth sesquisulfides are reasonably stable when exposed to air at room temperature, although a weak odor of hydrogen sulfide is frequently present. The orthorhombic A structure is found for La, Ce, Pr, Nd, Sm, Gd, Tb, and Dy. The monoclinic D structure is found for Dy, Ho, Er, and Tm (Dy_2S_3 is dimorphic), and the rhombohedral E structure is found for Yb and Lu. The B type of "rare-earth sesquisulfide" has been shown to be an oxysulfide,[2] and it is not obtained by this synthesis. The A or Th$_3$P$_4$ type of structure is also not obtained by this method. This form is usually (perhaps always) stabilized by impurities or nonstoichiometry.[1]

All crystals of the A and D structure types grow as rods (up to *ca.* 1 cm. in length), where the rod axis corresponds to the short crystallographic axis. Crystals of the E structure type grow as hexagonal plates which are generally twinned.

The crystals are electrically semiconducting.[1]

References

1. A. W. Sleight and C. T. Prewitt, *Inorg. Chem.,* 7, 2282 (1968).
2. D. Carre, P. Laruelle, and P. Besançon, *C. R. Acad. Sci. Paris,* 270, 537 (1970).

33. CADMIUM CHROMIUM(III) SELENIDE, CdCr$_2$Se$_4$

$$4Cd + 4Se + 2CrCl_3 \longrightarrow CdCr_2Se_4 + 3CdCl_2$$

Submitted by ARTHUR W. SLEIGHT*
Checked by HARRY L. PINCH†

Cadmium chromium(III) selenide can be prepared by the direct combination of the elements or by the reaction of cadmium selenide with chromium(III) selenide. Crystals of cadmium chromium(III) selenide have been prepared by flux growth[1] in cadmium chloride, by vapor transport,[2] and by a liquid-transport method[3] with a platinum metal "catalyst." The synthesis given

*Central Research Department, E. I. du Pont de Nemours & Company, Wilmington, Del. 19898.
†RCA Laboratories, Princeton, N.J. 08540.

here is simple, yet it gives a high yield of good quality crystals which are as large as any produced by the other methods.

Procedure

Although high-purity cadmium and selenium are commercially available, high-purity chromium(III) chloride is not. Therefore, if high-purity cadmium chromium(III) selenide is desired, it is best to make chromium(III) chloride* by passing chlorine over high-purity chromium metal at high temperatures (*ca.* 1200°C.).

Cadmium metal (4.0464 g., 0.036 mole), selenium (2.8426 g., 0.036 mole), and chromium(III) chloride (2.8505 g., 0.018 mole) are placed in a 15-mm. silica tube which has been sealed at one end. The tube is evacuated and sealed off to produce an ampul which is about 22 cm. long.

The ampul is heated slowly† to 900°C. in a muffle furnace and held there for 10 hours. ■ *Caution.* *The ampul should be heated in a hood! Toxic vapors will be produced if the ampul breaks.* The furnace is then cooled slowly to 500°C. over 7 days at a nearly linear rate. After this, the furnace is turned off and allowed to cool to room temperature.

The ampul is removed from the furnace and opened. ■ *Caution. Hydrogen selenide vapors are formed during the washing, and it is advisable that this operation also be performed in a hood.* The crystals of cadmium chromium(III) selenide can be washed in water to remove the cadmium chloride. The crystals are black octahedra which vary in size up to about 3 mm. on an edge. The only impurities detected in these crystals by arc emission spectrographic analysis were Mg and Cu, and these are present only in amounts of less than 20 p.p.m. *Anal.* Calcd. for $CdCr_2Se_4$: Cd, 21.1; Cr, 19.5; Se, 59.3. Found: Cd, 20.8; Cr, 17.7; Se, 58.6.

---

*Procedures for chromium(III) chloride are also given in *Inorganic Syntheses*, 2, 193; 5, 154; 6, 129.

†The checker found that heating rates greater than about 40°C./hour lead to explosions.

Properties

Cadmium chromium(III) selenide is a ferromagnetic semi-conductor with a Curie temperature of 130 K. Crystals grown by this method are p type. Cadmium chromium(III) selenide has the spinel structure with a cubic cell edge of 10.75 A.

References

1. G. Harbeke and H. Pinch, *Phys. Rev. Letters,* 17, 1090 (1966).
2. F. H. Wehmeier, *J. Cryst. Growth,* 5, 26 (1969).
3. H. von Philipsborn, *J. Appl. Phys.,* 38, 955 (1967).

34. GROWTH OF COBALT DISULFIDE SINGLE CRYSTALS

$$CoS_2\,(s) + Cl_2\,(g) \rightleftharpoons CoCl_2\,(g) + S_2\,(g)$$

Submitted by R. J. BOUCHARD*
Checked by R. KERSHAW† and A. WOLD†

Cobalt disulfide, an interesting member of the pyrite family, is unusual because it is ferromagnetic and metallic. A pure polycrystalline sample is difficult to prepare, requiring long reaction times, regrinding, and refiring. To measure electrical transport and magnetic and optical properties, large single crystals are desirable. The wide applicability[1] of chemical transport by transient volatile species suggested a technique for growing CoS_2 single crystals. A transport technique has additional advantages: It operates at moderate temperatures, which minimizes thermal defects; and it utilizes only small quantities of foreign components, which

*Central Research Department, E. I. du Pont de Nemours & Company, Wilmington, Del. 19898.
†Brown University, Providence, R.I. 02912.

reduces impurity substitution or occlusion that often attends single-crystal growth by flux techniques. The applicability of halogens in transporting pyrites has been demonstrated[2] in the preparation of small CoS_2 crystals by iodine.

Although crystals of cobalt disulfide can be obtained over a large range of temperatures and gradients using chlorine as a transporting agent, the larger crystals usually have holes in most of the crystal faces. Only the cycled reaction to be described gives crystals of high perfection in the 3–4-mm. range. The advantages of cycled transport reactions have been discussed by Scholz and Kluckow.[3] By this procedure imperfections are removed during the reverse part of the cycle, and thereby surface faults are healed. In addition, nucleation of new crystals is reduced, since only those crystals beyond a certain critical size survive the reverse gradient and continue to grow. This permits a large crystal to be grown without the problems of intergrowth with nearby growing crystals.

Procedure

Polycrystalline cobalt disulfide starting material is prepared by reacting the elements at 600°C. Stoichiometric quantities are placed in a silica tube, which is sealed under vacuum, then fired slowly to 600°C. in a muffle furnace for 3–4 days, and cooled slowly. The product is ground with an agate mortar and pestle under nitrogen for one hour. Then 10% excess sulfur is added, and the tube is sealed and fired as before. This process of grinding and firing usually has to be repeated to obtain a single-phase, stoichiometric product. The excess sulfur is removed by several washings with carbon disulfide (in a hood). A silica tube, 25 cm. long with 1 cm. i.d., is charged with approximately 0.5 g. (0.004 mole) of the polycrystalline CoS_2 and 0.01 g. of sulfur, which is added to drive the reaction

$$CoS_2\,(s) \rightleftharpoons CoS_{1+x}\,(s) + (1-x)\,S(g)$$

to the left. The tube is evacuated to approximately 10^{-3} mm., back-filled with 0.2 atmosphere of chlorine, and sealed off at a length of 15 cm. The crystal-growth end is thoroughly heated with a torch to eliminate any stray nuclei from the polycrystalline charge.

The sealed tube is then placed in a horizontal clamshell type of three-zone furnace.* A 45-mm.-i.d. silica tube, resting on collars of fire brick, is used as the effective furnace chamber to minimize the sharp boundary between zones. Temperatures in the center zone and two end zones are set at 900 and 885°C., respectively. After 24 hours, there is no detectable variation within ±1°C., as monitored by a Pt–Pt–10%Rh thermocouple. A fairly smooth, if not linear, gradient results between the ends of the silica tube. For one week, the temperature of the "growth end" of the sealed reaction tube is periodically cycled by varying its position from the hot (900°C.) central zone and the cool (885°C.) end zones of the furnace. This is conveniently accomplished by slowly pushing the tube from the end of the furnace with a piece of silica tubing. The ratio of time in the hot and cool zones is kept at 1:2, with one 8-hour period on a reverse-transport (*clean*) cycle and one 16-hour period on a forward-transport (*deposit*) cycle per day. During the second week, transport is allowed to proceed in the usual manner without any cycling. Only one crystal was produced at the cool tip. It was *ca.* 3–4 mm. on an edge and has almost perfect crystal faces. X-ray back-reflection data show it to have a cubic cobalt disulfide pyrite structure. The crystal was not analyzed but similar runs (not cycled) always yield crystals whose lattice constants and sulfur content are identical with the starting material, within experimental error.

Properties

Cobalt disulfide has a cubic pyrite structure, $a_0 = 5.5362(5)$A. It is ferromagnetic with a T_c of 124 K and shows metallic behavior from 4 K to room temperature.

---

*Lindberg Hevi-Duty furnace model 59000, Lindberg Hevi-Duty Company, 3709 Westchester Pike, Newton Square, Pa. 19073.

References

1. H. Schäfer, "Chemical Transport Reactions," Academic Press, Inc., New York, 1964.
2. B. Morris, V. Johnson, and A. Wold, *J. Phys. Chem. Solids,* 28, 1565 (1967).
3. H. Scholz and R. Kluckow, in "Crystal Growth," H. S. Peiser (ed.), p. 475, Pergamon Press, New York, 1967.

35. CHALCOGENIDE HALIDES OF COPPER, GOLD, MERCURY, ANTIMONY, AND BISMUTH

Submitted by A. RABENAU* and H. RAU†
Checked by R. KERSHAW‡ and A. WOLD‡

Most of these ternary compounds can, in principle, be prepared by high-temperature reactions, e.g., heating the respective elements or binary components (or both) together in sealed glass or silica ampuls. A certain knowledge of the thermal stabilities of the respective compounds is required. Separation from other phases often causes difficulties.

Hydrothermal synthesis in hydrogen halide acids leads directly to isolated single crystals of the ternary compounds suitable for x-ray investigation and physical measurement.[1,2]

No systematic study has been made to complete the list of substances given below. The reader is invited to look for further examples. The chalcogenide halides are formed by reaction of mixtures of the respective elements or binary components, or both, in the appropriate hydrogen halide acid under hydrothermal conditions. Temperatures are in the range of 500–100°C. The densities of the fluid are at least 55% of room-temperature values; the corresponding pressures are not known in most cases.

All substances described are prepared by the same technique, with slight variations of parameters, which are given separately.

*Max-Planck-Institut für Festkörperforschung, Stuttgart, Germany.
†Philips Research Laboratory, Aachen, Germany.
‡Brown University, Providence, R.I. 02912.

The procedure is described in detail, therefore, for one example, antimony(III) iodide sulfide.

A. ANTIMONY(III) IODIDE SULFIDE

$$Sb_2S_3 + 2HI \longrightarrow 2SbSI + H_2S$$

The conditions for the preparation of this compound are given in Table VI, Sec. F.

Procedure

The volume of the silica ampul which serves as the reaction vessel (Fig. 17) is measured, to the constriction, with a graduated buret. After subtracting the volume of the starting materials from the total volume of the ampul, the degree of fill can be determined (see Tables II–IV).*

Ten grams (0.03 mole) of antimony(III) sulfide is poured into the ampul with the aid of a suitable funnel. After the inside of the constriction has been cleaned with a cotton plug, the ampul is connected to a vacuum system, evacuated, and sealed off. During this operation the level of the liquid nitrogen should be corrected from time to time to avoid thawing of the acid, which would immediately react with the antimony(III) sulfide.

The sealed ampul is transferred from the liquid nitrogen and is placed immediately under running hot water until a film of liquid acid has formed on the inner wall of the ampul. In this way bursting of the ampul due to the increase in volume of the melting acid is avoided. The ampul is then kept in a hood until the acid is completely liquid. ■ *Caution. A heavy face shield should be worn during this operation.*

*For example, if the volume of the ampul is approximately 11.85 ml. and the volume of 10 g. (0.03 mole) antimony(III) sulfide is about 2.15 ml., then the free volume is 9.70 ml. Sixty-five percent of the free volume (6.3 ml. in this example) is filled with 10 M HI, and the ampul is dipped into liquid nitrogen almost up to the constriction to freeze the acid.

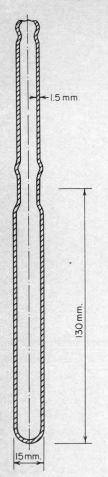

Fig. 17. Quartz glass ampul.

The ampul is inserted into an autoclave (Fig. 18), and the free volume remaining in the bore hole is calibrated with water from a buret, the water being removed afterward. Under experimental conditions, pressure is developed inside the ampul; to counterbalance this pressure, carbon dioxide is used. The amount of carbon dioxide is determined from the graph (Fig. 19), which takes into account the free volume in the autoclave, the temperature, and the pressure of experimental conditions. Because the internal pressures generated are not known in most cases, a very high external pressure is applied to prevent an explosion. As a result of the ampul design, reasonably high pressure can be tolerated. For safety reasons, however, 0.75–0.8 g. of carbon dioxide per milliliter of the free volume of the autoclave should be used. All experiments described have been performed under the following conditions: A rod of Dry Ice (solid carbon dioxide) of suitable diameter is prepared by keeping a piece of Dry Ice of sufficient size in the mold (Fig. 20), which is compressed between the jaws of a vise. A piece is cut off the rod with a knife, the weight being a little bit more than that calculated (plus 0.5 g. to counterbalance the evaporation losses during the closing procedure). The rod is kept on a balance until the correct weight is obtained. The piece is transferred immediately into the autoclave above the ampul. The locking cone is mounted and the screw cup bolted, with elongated (about 1-m.) wrenches. The force applied should be sufficient to deform the locking cone a little bit in the sealing region in order to get the autoclave gastight. ■ *Caution. During all manipulations, a heavy face shield should be used.*

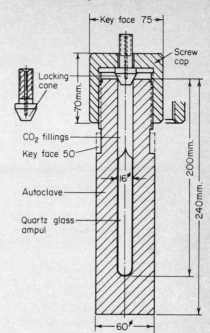

Fig. 18. Autoclave; material 24 CrMoV 55: ca. 0.45 Mn, 1.35 Cr, 0.55 Mo, 0.20 (for example, CV 120 of the Deutsche Edelstahlwerke).

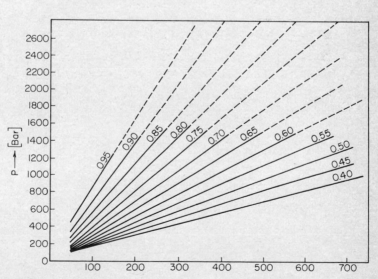

Fig. 19.   p-T diagram for carbon dioxide after Kennedy.[14] Parameters, g. CO_2/cc.

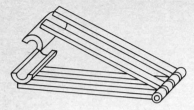

Fig. 20. Mold for pressing rods from Dry Ice.

The autoclave is mounted in a two-zone tube furnace. The axis of the furnace should be at an angle of about 15° to the horizontal, to cause high convection when a temperature gradient is applied.

If a temperature gradient is applied, the lower part (charge zone) is at the higher temperature. Temperatures are measured by thermocouples which are placed at the ends of the autoclave.

The furnace is subjected to a temperature/time treatment according to the specifications given in the tables for the respective compound. For the preparation of antimony(III) iodide sulfide, the furnace is cooled uniformly from 480–250°C. for a period of 10 days.*

The autoclave is then allowed to cool to room temperature. The screw cup is loosened carefully until carbon dioxide escapes.

■ *Caution. Because high pressure may exist inside the ampul (hydrogen sulfide, or in some cases, hydrogen), the following operations should be done in a hood behind a protective shield of Plexiglass or some other suitable material. Heavy, protective gloves should also be worn.*

The cup is removed from the autoclave, and the ampul is transferred into a Dewar vessel filled with liquid nitrogen. The cooled ampul is rolled in several layers of filter paper and opened by tapping with a hammer. This should be done with extreme caution. The contents of the ampul are transferred to a porcelain dish, washed several times with methanol, and dried between

---

*The ampul used by the checkers was 9 mm. i.d. X 130 mm. long. The charge consisted of 5 g. (0.015 mole) of antimony(III) sulfide and 4 ml. of 10 M HI. The charge temperature was 500°C., and the growth temperature was 285°C. Crystals of antimony(III) iodide sulfide up to 5 mm. long were grown in 24 hours.

layers of filter paper. The crystals of the respective compound can be isolated mechanically from accompanying material, e.g., silica chips or other solid phases.

The autoclave is ready for further experiments without any machining.

Notes on the Use of Tables II to VI

The data given in the following tables refer to a silica ampul of the size shown in Fig. 17. The inner volume of the ampuls will be between 10 and 15 ml.

The headings of the columns in Tables II–VI are given as follows:

1. Column 1 lists the compounds to be prepared.

2. Column 2 indicates the starting materials, which are usually analytic grade and are added as powders. The purity of the crystals depends upon the starting materials, although increased purification has been observed, as a result of the reaction.

3. For the acids listed in column 3, the densities used were 1.19 (12 M HCl), 1.50 (9 M HBr), 1.95 (10 M HI).

4. The term *degree of fill* (column 4) is explained in the general procedure section.

5. The two temperatures shown in Table III, for example, 350°C. → 150°C. means the autoclave is initially heated to 350°C. and is then uniformly cooled to 150°C. within the time given in column 6. In Table IV, the fifth column lists the temperature, for example, 450:430°C., which indicate that the autoclave is heated to 450°C. at the lower end (charge zone) and 430°C. at the upper end, and is kept under these conditions for the time given in column 6. After being heated in this manner, the autoclave is cooled to room temperature. In some cases a cooling program follows the heat treatment.

6. The ampul normally does not contain the elements in stoichiometric proportions indicated by the compositions given in column 1. Moreover, an isothermal reaction does not take place.

TABLE I

Compound	Thermal stability,* °C.	Symmetry	Unit cell dimensions					Possible space group
			a, A.	b, A.	c, A.	$\beta°$	z	
$CuITe_2$	400	Monoclinic	8.672	4.881	16.493	135.0	4	$P\,\dfrac{2_1}{c}$
$CuBrTe_2$	416		8.358	4.951	15.704	135.1	4	
$CuClTe_2$	ca. 415		8.168	4.195	15.187	134.9	4	
$CuITe$	442	Tetragonal	17.08		4.87		16	$I\,\dfrac{4_1}{amd}$
$CuBrTe$	~440†		16.40		4.72			
$CuClTe$	400		15.63		4.78			
$CuISe_3$	394	Rhombohedral‡	14.083		14.187		18	R3, R$\bar{3}$, R32 R3m, R$\bar{3}$$m$
$CuBrSe_3$	338	Orthorhombic	14.344	7.678	4.469		4	$P\,nc2$, $P\,mna$
$CuClSe_2$	319	Monoclinic§	7.735	4.665	30.89	90.4	12	$P\,2$, $P\,m$, $P\,\dfrac{2}{m}$

*In the absence of air; melting or decomposition point.
†Phase transformation at 280° C.
‡Referred to hexagonal cell.
§Pseudorhombohedral.

TABLE II

Compound (1)	Starting materials (2)	Solvent (3)	Degree of fill, % (4)	Temperature, °C. (5)	Time, days (6)	Remarks (7)
$CuClSe_2$	15 g. CuCl 5 g. Se	12 M HCl	55	$350 \rightarrow 150$	10	Few black needles; some Se, much CuCl
$CuBrSe_3$	15 g. CuBr 5 g. Se	9 M HBr	65	$340 \rightarrow 150$	10	Dark red crystals, about 1 mm.; much CuBr
$CuISe_3$	10 g. CuI 5 g. Se	5 M HI	60	$390 \rightarrow 200*$	10	Dark violet crystals, about 2.5 mm.; some CuI
$CuClTe$	20 g. CuCl 1 g. Te	1 M HCl	58	$350 \rightarrow 150$	10	Few black needles, $\emptyset 0.3$ mm., $l \leqslant 15$ mm.; much CuCl
$CuBrTe$	20 g. CuBr 1 g. Te	1 M HBr	60	$400 \rightarrow 150$	10	Black needles, $\emptyset 0.5$ mm., $l \leqslant 15$ mm.; much CuBr
$CuTe$	10 g. Cu_4Te_3	10 M HI	65	$500 \rightarrow 200$	10	Grayish, interlocked needles, $\emptyset 1$ mm., $l \leqslant 15$ mm.
$CuClTe_2$	15 g. CuCl 5 g. Te	12 M HCl	55	$350 \rightarrow 150$	10	Thin, black needles up to 20 mm.; much CuCl
$CuBrTe_2$	15 g. CuBr 5 g. Te	9 M HBr	65	$400 \rightarrow 150$	10	V. small black needles; much CuBr
$CuITe_2$	15 g. CuI 1 g. Te	1 M HI	60	$440 \rightarrow 150$	10	Black rhombic crystals, 0.5 mm.

*Autoclave in horizontal position.

TABLE III

Compound (1)	Starting materials (2)	Solvent (3)	Degree of fill, % (4)	Temperature, °C. (5)	Time, days (6)	Remarks (7)
AuTe*	6.2 g, Au† 4 g, Te 8 g, I	10 M HI	65	450 → 150	10	Compact grayish crystals, 2–3 mm. Au
AuTe$_2$	6.2 g, Au† 4 g, Te	10 M HI	65	450 → 150	10	Silvery white spears ∅ 1 × 0.5 mm. l ≤ 50 mm., Au
AuBrTe$_2$	13.9 g, Au† 5.1 g, Te 5.7 g, Br	9 M HBr	65	350 → 150	10	Silvery white, square crystals 3 × 4 × 0.5 mm., Au
AuClTe$_2$	5.9 g, Au† 1.9 g, Te 1.0 g, Cl	HCl‡	0.02 mole/cc.	400 → 100	10	Silvery white, square crystals, Au

*The corresponding AuBrTe and AuClTe do not seem to exist.
†Gold powder, precipitated from solution with SO$_2$.
‡Condensed into the ampul via a vacuum system.

TABLE IV

Compound (1)	Starting materials (2)	Solvent (3)	Degree of fill, % (4)	Temperature, °C. (5)	Time, days (6)	Remarks (7)
γ-Hg$_3$S$_2$Cl$_2$	27 g. HgS	12 M HCl	55	450:430	10	Amber crystals up to 5 mm.

TABLE V

Compound (1)	Starting materials (2)	Solvent (3)	Degree of fill, % (4)	Temperature, °C. (5)	Time, days (6)	Remarks (7)
$Pb_7S_2Br_{10}$	7.2 g. PbS 5.5 g. $PbBr_2$	9 M HBr	65	360:340*	10	Thin, orange needles $\leq$ 4 mm., some PbS
$Pb_5S_2I_6$	10 g. PbS	10 M HI	65	480:465 480 $\rightarrow$ 100	6 10	Carmine needles $\emptyset$ 1.5 mm., l $\leq$ 50 mm.
Pb_4SeBr_6	5 g. PbSe 20 g. $PbBr_2$	9 M HBr	65	350:335 350 $\rightarrow$ 200	5 5	Nearly black compact crystals, much $PbBr_2$, some PbSe and Se

*Autoclave in horizontal position

TABLE VI

Compound (1)	Starting materials (2)	Solvent (3)	Degree of fill, % (4)	Temperature, °C. (5)	Time, days (6)	Remarks (7)
$Sb_4S_5Cl_2$	10 g. Sb_2S_3	12 M HCl	55	490 $\rightarrow$ 140	10	Gray crystals up to 5 mm.
SbSBr	10 g. Sb_2S_3	9 M HBr	65	455:435	7	Orange needles, $\emptyset$ 0.3 mm., l $\leq$ 5 mm., some Sb_2S_3
SbSI	10 g. Sb_2S_3	10 M HI	65	490 $\rightarrow$ 250	10	Dark red needles, l $\leq$ 50 mm.
BiSBr	10 g. Bi_2S_3	9 M HBr	65	500 $\rightarrow$ 200	10	Carmine needles 0.5 × 0.5 × 30 mm.
BiSI	10 g. Bi_2S_3	10 M HI	65	500 $\rightarrow$ 200	10	Black needles with metallic luster, $\emptyset$ 0.7 mm. l $\leq$ 25 mm.

Therefore, crystals of various phases may be obtained at the same time.

The conditions given have been found empirically to be satisfactory. No attempts have been made to optimize the data.

B. COPPER CHALCOGENIDE HALIDES[3,4]

Copper chalcogenide halides have been prepared by this method for the first time.* Tables I and II give the data. The conditions for the preparation of these compounds are listed in Table II.

Properties

The crystals are stable in air. The tellurium compounds are stable in alkaline solutions but decompose in concentrated nitric and sulfuric acid. The selenium compounds decompose readily in alkaline solutions.

All compounds exhibit a temperature-independent diamagnetism which suggests copper to be monovalent.

The phases form part of pseudobinary systems between the copper(I) halide and selenium or tellurium, respectively.

The optical band gaps at room temperature from spectral reflectance measurements[5] are 1.4 e.V. for the CuTeX compounds, 1.2 e.V. for the $CuTe_2X$ compounds, 1.6 e.V. for $CuSe_2Cl$, and about 2.0 e.V. for $CuSe_3Br$ and $CuSe_3I$. Thermal stabilities and crystallographic data are summarized in Table I.

C. GOLD HALIDE TELLURIDES[6,7]

Gold halide tellurides have been prepared by this method for the first time. The conditions for the preparation of these compounds are listed in Table III.

---

*The corresponding silver compounds do not seem to exist.

Properties

The compounds are insoluble in dilute acids and alkalis. They decompose in dilute nitric and concentrated sulfuric acids. They decompose or melt in the absence of air at (AuTeI) 360°C., (AuTe$_2$ I) 440°C., (AuTe$_2$ Br) 475°C., and (AuTe$_2$ Cl) 447°C. AuTeI crystals are monoclinic with cell parameters a = 7.245 A., b = 7.622 A., c = 7.313 A., β = 106.3°. AuTe$_2$ I, AuTe$_2$ Br, and AuTe$_2$ Cl are orthorhombic with a = 4.735 A., b = 4.064 A., c = 12.55 A.; a = 8.946 A., b = 4.038 A., c = 12.389 A.; and a = 8.777 A., b = 4.021 A., c = 11.873 A. AuTe$_2$ I, AuTe$_2$ Br, and AuTe$_2$ Cl all exhibit metallic conductivity.

D. MERCURY(II) HALIDE SULFIDE

The mercury chloride sulfide γ-Hg$_3$ S$_2$ Cl$_2$ is obtained by either quenching a mixture of mercury(II) sulfide vapor from 750°C. or by the reaction of a dilute alkaline solution of mercury(II) chloride with carbon disulfide. The conditions for the preparation of Hg$_3$ S$_2$ Cl$_2$ are given in Table IV.

Properties

The γ modification has been found to be unstable under normal conditions.[8] The light-yellow crystals decompose slowly within months. Crystallographic data for the orthorhombic compounds are a = 9.09$_4$ A., b = 16.84$_3$ A., c = 9.34$_9$ A., and z = 8.

E. LEAD(II) CHALCOGENIDE HALIDES[9]

Various lead halide sulfides are reported in the literature. They precipitate when acidic solutions of lead halides react with hydrogen sulfide. Most probably the lead(II) chalcogenide halides described below are the only stable ones among these compounds. The conditions for the syntheses of these materials are given in Table V.

The compounds decompose in acids or alkalis and precipitate black lead sulfide. The stability decreases in the sequence $Pb_5S_2I_6$, Pb_4SeBr_6, $Pb_7S_2Br_{10}$. The compounds belong to a pseudobinary system, formed between lead chalcogenide and the respective lead(II) halide. The crystals melt or decompose in the absence of air: $Pb_5S_2I_6$, 418; $Pb_7S_2Br_{10}$, 381; and Pb_4SeBr_6, 370°C. From reflectance measurements, the optical bandgaps at room temperature were found to be about 1.6 e.V. for Pb_4SeBr_6, and 2.0 e.V. for both $Pb_5S_2I_6$ and $Pb_7S_2Br_{10}$.[10]

The crystallographic data for these compounds are monoclinic $Pb_5S_2I_6$: a = 14.33 A., b = 4.434 A., c = 14.53 A., β = 98.0°; hexagonal $Pb_7S_2Br_{10}$: a = 12.27 A., c = 4.318 A.; orthorhombic Pb_4SeBr_{10}: a = 4.36 A., b = 9.72 A., c = 15.78 A.

F.   ANTIMONY AND BISMUTH HALIDE SULFIDES

The compounds have been prepared by annealing stoichiometric amounts of the elements or the respective antimony(III) or bismuth(III) sulfide and halides in sealed silica ampuls.[11,12] Good crystals are obtained by using a surplus of M_2X_3.[12] The conditions for the preparation of these thiohalides are given in Table VI.

Properties

The crystals are stable in air and dilute acids, but are attacked by strong acids. Needles cleave along the needle axes. The compounds are orthorhombic [11] (space group D_{2h}^{16}), $Sb_4S_5Cl_2$: a = 10.5 A., b = 11.1 A., c = 9.4 A.; SbSBr: a = 8.2 A., b = 9.7 A., c = 4.0 A.; SbSI: a = 8.5 A., b = 10.1 A., c = 4.2 A.; BiSBr: a = 8.1 A., b = 9.7 A., c = 4.0 A.; BiSI: a = 8.5 A., b = 10.2 A., c = 4.1 A.

The compounds have been of interest because of their physical properties. They are both photoconducting and ferroelectric. Antimony(III) iodide sulfide has a Curie point at 22°C.[12]

References

1. H. Rau and A. Rabenau, *Mater. Res. Bull.*, 2, 609 (1967).
2. H. Rau and A. Rabenau, *Solid State Commun.*, 5, 331 (1967).
3. A. Rabenau, H. Rau, and G. Rosenstein, *Naturwiss.*, 56, 137 (1969).
4. A. Rabenau, H. Rau, and G. Rosenstein, *Z. Anorg. Allgem. Chem.*, in press.
5. A. Rabenau and H. Rau, *Solid State Commun.*, 7, 1281 (1969).
6. A. Rabenau, H. Rau, and G. Rosenstein, *Angew. Chem.*, 81, 148 (1969); *Angew. Chem. Inter. Ed. Engl.*, 8, 145 (1969).
7. H. Rau and A. Rabenau, *III Int. Conf. Solid Compds. Transition Elems.*, June 16-20, *Oslo, Norway*, 1969.
8. H. Puff, A. Harpain, and K. P. Koop, *Naturwiss.*, 53, 274 (1966).
9. A. Rabenau and H. Rau, *Z. Anorg. Allgem. Chem.*, 369, 295 (1969).
10. A. Rabenau, H. Rau, and G. Rosenstein, *Naturwiss.*, 55, 82 (1968).
11. E. Dönges, *Z. Anorg. Allgem. Chem.*, 263, 112 (1950).
12. R. Nitsche and W. Merz, *J. Phys. Chem. Solids*, 13, 154 (1960).
13. E. Fatuzzo, G. Harbeke, W. J. Merz, R. Nitsche, H. Roetschi, and W. Ruppel, *Phys. Rev.*, 127, 2063 (1962).
14. G. C. Kennedy, *Am. J. Sci.*, 252, 225 (1954).

IV. PHOSPHIDES AND SILICIDES

36. PYRITE TYPE OF SILICON DIPHOSPHIDE

$$Si + 2X_2{}_{(g)} \longrightarrow SiX_4{}_{(g)}$$
$$SiX_4{}_{(g)} + \tfrac{1}{2}P_4{}_{(g)} \longrightarrow SiP_2 + 2X_2{}_{(g)}$$

Submitted by P. C. DONOHUE*
Checked by J. FERRETTI† and A. WOLD†

The synthesis of black, pyrite type of silicon diphosphide was first accomplished at high pressure[1] (15-50 kb.). It has since been prepared by vapor-transport techniques.[2,3] A low-pressure, red form of SiP_2 is produced by sublimation from 900-500°C. in a

*Central Research Department, E. I. du Pont de Nemours & Company, Wilmington, Del. 19898.
†Brown University, Providence, R.I. 02912.

tube in the absence of a transporting agent.[4] At a region of about 600°C., small quantities of the black pyrite type of phase condense. In the presence of a halogen transporting agent, large yields of only the pyrite type of phase are formed.

The chemical bonding in the two forms is quite different. The low-pressure form has the $GeAs_2$ type of structure, in which the coordination of the atoms is normal, i.e., fourfold for Ge and threefold for As. In the pyrite type silicon is octahedrally coordinated, and phosphorus is tetrahedrally coordinated. The difference is reflected in the density and properties. The density of the $GeAs_2$ type is 2.47 g./cc., whereas for the pyrite type it is 3.22 g./cc. The pyrite type is a metallic conductor, whereas the $GeAs_2$ type is a semiconductor.

Procedure

The pyrite type of SiP_2 is prepared, with halogen transport techniques, by reaction of the elements in an evacuated silica tube. Highly purified silicon, phosphorus (red form), and halogens may be obtained commercially.* Chlorine, bromine, or iodine may be used as the transporting agent. The reaction is carried out in a tube furnace having at least two independent zones capable of reaching 1200°C. A tube furnace having a natural gradient may also be used provided a temperature profile is made, and the tube containing the reactants can be positioned so as to obtain the desired temperatures at its hot and cooler ends.

Reactants should be weighed in the atomic ratio of 1Si/ 2.0–2.5P. The presence of excess phosphorus has been found to enhance reaction and impede the formation of silicon monophosphide. Since a large excess of phosphorus increases the chance of explosion from its vapor pressure, an optimum ratio is

*Electronics Space Products, Inc., Los Angeles, Calif. 90000; Research Organic Inorganic Chemical Corporation, Sun Valley, Calif. 91352.

1Si/2.1P. To decrease further the possibility of explosion, the reaction tube should be heated slowly (over a period of 48 hours) to reaction conditions. The furnace should be well shielded in case of explosion. In order to speed reaction, the silicon should be ground to about a 100–200 mesh and introduced along with the phosphorus lumps into a thoroughly cleaned silica tube, preferably heavy walled (16 mm. o.d., 10 mm. i.d.). Enough halogen must be added to the tube to produce a vapor pressure of approximately 2–4 atmospheres at operating conditions. Bromine or iodine may be used most conveniently.

The tip of the tube containing the reactants should be placed in ice water, and a weighed amount of iodine or bromine (2–3 drops) added. To avoid condensation of water vapor, the tube should be transferred quickly to a vacuum line and promptly evacuated while the tip is maintained at ice-water temperature. The tube is sealed finally at a length of about 7 in.

Chlorine may be used as the transporting agent by evacuating the tube, introducing about 1 atmosphere of gas, and sealing the tube.

The sealed tube is placed in the tube furnace, and the temperature is raised slowly over a period of about 2 days. The end containing the reactants should be maintained at 1100–1200°C., and the other end at about 700°C. After 2 or 3 days in these conditions, black, shiny crystals up to 1 mm. on an edge form in the tube at a region of about 800°C. The tube may be cooled in the furnace by turning off the power. When the tube has cooled to room temperature, it may be removed and opened. Owing to the possible presence of excess phosphorus, the tube should be opened in a hood and the superfluous phosphorus allowed to burn off. In addition, the precaution of gloves and a face shield should be taken. The crystals are washed in water and acetone and allowed to dry. *Anal.* Calcd. for SiP_2: Si, 68.8; P, 31.2. Found: Si, 68.51; P, 31.33.

Properties

The pyrite type of SiP_2 forms as black, shiny crystals; the cubic cell dimension is $a = 5.7045 \pm 0.0003$ A. The material is stable hydrolytically and is thermally stable in air to 900°C. It is a good metallic conductor with room temperature resistivity of $\rho = 3 \times 10^{-5}$ Ω-cm.

References

1. J. Osugi, R. Namikawa, and Y. Tanaka, *J. Chem. Soc. Japan,* Pure Chem. Sec., **87,** 1169 (1966).
2. T. Wadsten, *Acta Chem. Scand.,* **21,** 1374 (1967).
3. P. C. Donohue, W. J. Siemons, and J. L. Gillson, *J. Phys. Chem. Solids,* **29,** 807 (1968).
4. T. Wadsten, *Acta Chem. Scand.,* **21,** 593 (1967).

37. SINGLE CRYSTALS OF IRON MONOPHOSPHIDE

$$3NaPO_3 + FeF_3 \xrightarrow[\text{current}]{\text{electric}} FeP + 3NaF + P_2O_5 + 2O_2$$
$$FeP + I_2 \longrightarrow FeI_2 + [P]$$

Submitted by DAVID W. BELLAVANCE* and AARON WOLD†
Checked by WALTER KUNNMANN‡

The various techniques for the synthesis of transition metal phosphides have been summarized in the monograph by Aronsson, Lundstrom, and Rundqvist.[1] The most common method used for the preparation of the phosphides is the direct combination of the elements. However, in general, the products are not pure or

*Texas Instruments Incorporated, P.O. Box 5936, Dallas, Tex. 75222.
†Brown University, Providence, R.I. 02912
‡Brookhaven National Laboratories, Upton, N.Y. 11973.

homogeneous, and any crystals produced are usually small and of poor quality. In addition, there are the problems of maintaining high-purity reduced metals and the handling of phosphorus in the elemental state. Andrieux and Chêne[2-8] first demonstrated that fused salt electrolysis could be used to prepare transition-metal phosphides. Wood,[9] Yocom,[10] and Hsu and coworkers[11,12] developed this technique further and reported the growth of small crystallites for many of these compounds. Each of these investigators electrolyzed a melt containing a transition-metal oxide and sodium metaphosphate. Sodium fluoride or carbonate was added to facilitate solution of the oxide. However, because of incomplete dissolution of the metal oxide in the phosphate melt and incomplete reduction of the metal complexes present in the melt, the products were often of uncertain composition and contaminated by transition-metal oxide. More recently Bellavance, Vlasse, Morris, and Wold[13] observed that the use of iron(III) fluoride in place of iron(III) oxide was preferable since it is readily soluble in the sodium metaphosphate, and relatively low reduction potentials (*ca.* 0.7 volt) are sufficient to give the desired product. The composition of the product obtained by electrolysis is dependent on the phosphorus/iron ratio in the fused melt and the reaction temperature. The quality and size of the crystals depend upon the applied current density and reaction time.

Large, homogeneous crystals of FeP can also be grown by the method of chemical transport.[13,14] In this technique a charge material reacts in a sealed tube with a transporting agent at elevated temperatures to form volatile compounds. The vapor then diffuses to a region in the tube of lower temperature where the reverse reaction is thermodynamically favored, and the material is deposited as a single crystal. Both iodine and chlorine have been used as the transporting agents.[14] However, it was observed that the nucleation rate with chlorine is difficult to control, and iodine is therefore preferred for the growth of large, single crystals.[14]

Procedure

The apparatus for the electrolysis of the fused melts is shown in Fig. 21. The main body, consisting of a mullite tube with water-cooled brass cap ends sealed by rubber O rings, provides an enclosed chamber for electrolysis. Gas connections at either end allow the reaction to be carried out under a flowing argon atmosphere. Exhaust gas is bubbled through a 1:1 solution of hydrochloric acid and formaldehyde before the gas passes to the open atmosphere. The crucible has a cavity 1-in. i.d. X 5-in. depth and is machined from a $1\frac{1}{2}$-in., high-purity carbon rod, Ultra-Carbon.* The crucible is centered in the hot zone of the furnace by a carbon pedestal. A centered $\frac{1}{4}$-in. carbon rod serves as the cathode, and the crucible acts as the anode during electrolysis. External cathode and anode current leads are attached to a $\frac{1}{4}$-in. stainless-steel support rod and the brass base plate, respectively. The use of internal wire connections is prohibited by the corrosive nature of the atmosphere in the chamber during electrolysis. The entire assembly is vertically mounted in a split-tube furnace. The

*United Carbon Products Company, Inc., Bay City, Mich. 48706.

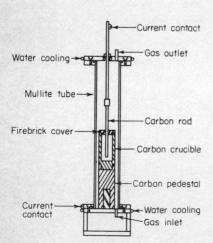

Fig. 21. *Assembly for fused-salt electrolysis.*

thermocouple for the control of the furnace temperature is placed outside the reaction chamber (mullite tube) and is calibrated with respect to a thermocouple placed within the reaction vessel in the absence of a melt.

Six grams (0.053 mole) of iron(III) fluoride* is mixed thoroughly with 60 g. (0.1 mole) of purified-grade sodium metaphosphate† in a beaker and the contents are transferred to the carbon crucible. The crucible is placed in the reaction chamber and is heated to 925°C. at the rate of 75°C./hour in a flowing argon atmosphere. The crucible is allowed to equilibrate at 925°C. for 1 hour. The center electrode is inserted to a depth of $\frac{3}{4}$ in. and a current of 250 mA. (61 mA./cm.^{2}) is used. A constant current source is used to maintain the desired current. The electrolysis should be carried out from 12 to 24 hours to give a sufficient yield. After electrolysis the melt is cooled to room temperature at the rate of 75°C./hour. The crucible is cut to expose the reaction boule, and the excess melt is removed by leaching in hot, dilute hydrochloric acid. Free carbon is removed by flotation with methylene iodide(diiodomethane). The product is in the form of metallic needles and small crystallites with a yield of 0.5–1 g.

The H tube for the preparation of the iodine transport tubes is shown in Fig. 22. The assembly is cleaned thoroughly in aqua regia and dried prior to use. A homogeneous powder of iron monophosphide which has been prepared electrolytically is used as the transport charge. One gram of the charge material is introduced into the reaction section (13-mm. silica tubing) with a long-stem funnel and the tube is sealed at point A. It is necessary to weigh out 100 mg. of iodine which has been resublimed previously and dried in a desiccator over phosphorus(V) oxide in a melting-point capillary. The capillary is evacuated to a pressure of 10^{-3} torr and is sealed with a torch. The iodine capillary is placed in the other side of the H tube, and the assembly is attached to a vacuum system at point B. The sample is evacuated to a pressure of 10^{-5}

*Ozark-Mahoning Co., 1870 S. Boulder Ave., Tulsa, Okla. 74119.
†Fischer Scientific Company, 711 Forbes Ave., Pittsburgh, Pa. 15219.

torr and is outgassed with slight heating for 2–3 hours. The assembly is removed from the vacuum system by sealing with a torch at point *B*. The iodine capillary is broken, being struck by the enclosed weight, and the iodine is sublimed onto the transport charge by submerging the charge in liquid nitrogen. The silica reaction tube* containing the transport charge and iodine is sealed off at a length of 8 in. The tube is placed in a Hevi-Duty MK 2012 furnace which has been rewired to give two independently controlled temperature zones. A 2-cm.-thick firebrick baffle placed between the two zones gives a more satisfactory temperature profile. The growth (empty) and charge zones are set at 800 and 500°C., respectively, for back transport. This minimizes the number of nucleation sites in the growth zone. After 1 day the temperatures of the zones are reset to 800°C. for the charge zone

---

*All sealed tubes containing chemical reagents are potential bombs and should be handled with appropriate shielding for hands, body, and face.

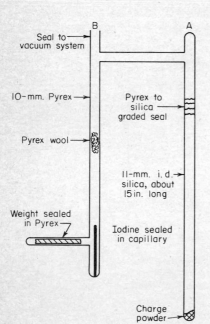

B   A

Seal to→
vacuum system

10-mm. Pyrex→   Pyrex to
                    silica →
                    graded seal

Pyrex wool→

                    11-mm. i.d.→
                    silica, about
                    15 in. long

Weight sealed
in Pyrex →   Iodine sealed
                    in capillary

                    *Fig. 22. H tube for preparation
                    of iodine transport tubes.*

Charge
powder→

and 500°C. for the growth zone. The transport is continued for 1–2 weeks. At the end of the transport period, the furnace is shut off and allowed to cool to room temperature. The crystals are removed from the tube and washed with carbon tetrachloride and acetone to remove excess iodine. The crystals may be either in the form of plates up to $3 \times 2 \times 1$ mm. or large blocks up to $3 \times 3 \times 3$ mm., all with well-defined faces.

Properties

Iron monophosphide grows in the form of needles by using fused-salt electrolysis, or in the form of plates and polyhedra when grown by the chemical transport technique. The crystalline faces are particularly well formed by the transport technique. All crystals show metallic luster. The x-ray powder pattern can be indexed on the basis of an orthorhombic unit cell with $a = 5.19$ A., $b = 3.099$ A., and $c = 5.79$ A. The crystals are antiferromagnetic ($T_N = 123$ K) and metallic. The material is stable in air, water, and mineral acids at room temperature. Iron content may be determined by the silver reductor method, by titrating with standardized cerium(IV) sulfate. The phosphorus content may be determined indirectly by treating the phosphide with vanadium(V) sulfate solution and reoxidizing the vanadium(IV) produced with a standard solution of potassium permanganate. *Anal.* Calcd. for FeP: Fe, 64.33; P, 35.67. Found: Fe, 64.17; P, 35.60.

References

1. B. Aronsson, T. Lundstrom, and S. Rundqvist, *Borides, Silicides, and Phosphides,* John Wiley & Sons, Inc., New York, 1965.
2. J. L. Andrieux, *Rev. Met.,* 45, 49 (1948).
3. J. L. Andrieux and M. Chêne, *Compt. Rend.,* 206, 661 (1938).
4. J. L. Andrieux and J. Chêne, *ibid.,* 1562 (1938).
5. J. L. Andrieux and J. Chêne, *ibid.,* 209, 672 (1939).
6. M. Chêne, *ibid.,* 207, 571 (1938).
7. M. Chêne, *ibid.,* 208, 1144 (1939).
8. M. Chêne, *Ann. Chim.,* 15, 266 (1941).
9. D. W. Wood, Ph.D. thesis, University of Illinois, Urbana, Ill., 1953.

10. P. N. Yocom, Ph.D. thesis, University of Illinois, Urbana, Ill., 1958.
11. S. S. Hsu, P. N. Yocom, and T. C. C. Cheng, *Interim Report,* contract N6-ori-071(50)-NR-52-341, December 1953–July 1955.
12. S. S. Hsu, P. N. Yocom, et al., *Final Report, ibid.,* July 1953–December 1958.
13. D. Bellavance, M. Vlasse, B. Morris, and A. Wold, *J. Solid-State Chem.,* 1, 82 (1969).
14. D. Bellavance, Ph.D. thesis, Brown University, Providence, R.I., 1970.

38. COPPER-SOLVENT FLUX GROWTH OF MANGANESE MONOSILICIDE

Submitted by VANCLIFF JOHNSON*
Checked by ARLINGTON FINLEY† and AARON WOLD†

Silicide crystals are sometimes grown conveniently from metal fluxes when growth by conventional melt techniques is problematic. Jangg, Kieffer, and Kögler[1] were able to grow crystals of high-melting (>2000°C.) tungsten and molybdenum silicides by slowly cooling Cu/W, Mo/Si solutions from 1300°C. Because of the high vapor pressure of manganese at the melting point of MnSi (1275°C.), manganese losses are unavoidable in melt growth. It is therefore desirable to grow MnSi by other methods, as for example, from metal solvents.

A suitable metal solvent should provide high solubility for the reactants. In addition, it should either not form compounds with the reactants or only those with heats of formation that are low with respect to the desired silicide. Finally, the solvent should have a relatively low melting point and vapor pressure, and separate easily from the product silicide.

Copper metal is a good flux for manganese silicides, since both manganese and silicon are highly soluble in copper. Manganese silicides are the most stable phases in the Cu–Mn–Si system at

*Central Research Department, E. I. du Pont de Nemours & Company, Wilmington, Del. 19898.
†Department of Chemistry, Brown University, Providence, R.I. 02912.

most compositions, and the residual copper-rich solid solution can be dissolved easily in dilute nitric acid. However, this method has the disadvantage of product contamination from either flux inclusion or chemical substitution.

Procedure

Two grams (0.036 g. atom) of manganese, 2 g. (0.071 g. atom) of silicon, and 10 g. (0.16 g. atom) of copper are placed as pieces in a 10-ml. Al_2O_3 crucible (Morganite*) in a silica ampul with an open end. After evacuation and sealing, the ampul is placed in a Globar muffle furnace. The temperature is raised to 1200°C. over a period of 12 hours, then lowered to 500°C. at 10°C./hour. The ampul is removed, allowed to cool, inserted in a piece of rubber tubing of larger diameter, and broken in order to obtain the ingot safely. The copper-rich Cu–Mn–Si matrix is dissolved in 8 N nitric acid, which leaves octahedral crystals of MnSi, and these are then washed with water and dried. Several crystals of up to 2 mm. on edge are obtained.

Characterization

Several octahedral crystals were ground to powder and studied by the Guinier-Hägg technique (Cu radiation). All diffraction lines could be indexed on the basis of a cubic cell with a = 4.560 ± 0.001 A., which is within experimental error of that reported previously for MnSi, 4.558 ± 0.001 A.[2] In view of the similarity of the cell constants of the ground crystals to those of previous preparations, *significant* replacement of manganese by copper is not indicated. This was checked on a few crystals by emission spectrographic analysis which indicated an upper limit of *ca*. 1% of copper.

It should be noted that the starting atomic ratio of manganese to silicon is not 1:1, but 1:1.96. An excess of silicon is necessary,

*Morganite, Inc., 3302-3320 48th Ave., Long Island City, N.Y. 11100.

since the activity of silicon is much less than that of manganese in the molten system. Jangg, Kieffer, and Kögler[1] have discussed this generally. The exact ratio is probably not critical, and MnSi may be expected in a large range of Mn/Si ratios, as found for tungsten and molybdenum silicides.[1] It should also be possible to prepare crystals of Mn_3Si, Mn_5Si_3, and the defect silicides close to Mn_4Si_7 from copper by varying the Mn/Si ratio.

References

1. G. Jangg, R. Kieffer, and H. Kögler, *Z. Metallkd.*, 59, 546 (1968).
2. W. B. Pearson, "Handbook of Lattice Spacings and Structures of Metals," Vol. 2, p. 317, Pergamon Press, New York, 1967.

V. HYDRIDES

39. SINGLE-CRYSTAL CERIUM HYDRIDES

$$M + \frac{s}{2}H_2\,(g) \longrightarrow MH_s\,(s) \qquad 3 \geqslant s \geqslant 1.8$$

Submitted by GEORGE G. LIBOWITZ* and JOHN G. PACK*
Checked by C. E. LUNDIN†

All the metallic transition-metal hydrides can be prepared by direct reaction at elevated temperatures. The method has been used commonly to prepare polycrystalline material according to the exothermic reaction described by the above equation.

The dihydrides of the rare-earth metals are an unusual group of compounds with respect to defect structure, bonding mechanism, and electronic properties.[1] These compounds usually show wide deviation from stoichiometry. In some individual cases the

*Ledgemont Laboratory, Kennecott Copper Corporation, Lexington, Mass. 02173.
†Denver Research Institute, University Park, Denver, Colo. 80210.

hydrogen/metal ratio varies from 1.8 to 3. Prior studies[2] made on polycrystalline hydrides have shown that in order to elucidate their inherent electronic properties, it is essential to work with single-crystal samples.

Single crystals of transition-metal hydrides cannot be grown from the melt by the usual techniques because the hydrides dissociate at temperatures well below their melting points. The method used here is one of compound formation from saturated elemental melts. It is a modification of the technique used by Harman et al.[3] and Stambaugh et al.[4] for growing crystals of III–V semiconducting compounds.

The principle of the method can be seen by referring to Fig. 23, which is a partial phase diagram of the cerium hydrogen system. Molten cerium metal is placed within a thermal gradient such that the lowest temperature is above the peritectic temperature (*ca.* 1010°C., from the diagram), and hydrogen is slowly dissolved in

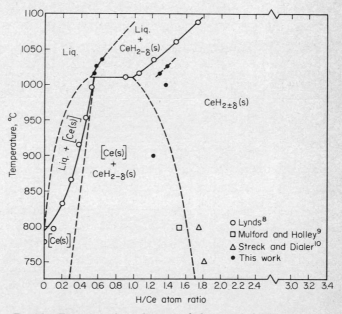

Fig. 23. Tentative phase diagram of the cerium hydrogen system at elevated temperatures.

the molten metal. It can be seen from Fig. 23 that when that part of the melt which is at the lowest temperature attains the hydrogen content corresponding to the liquidus line between the liquid and the two-phase (liquid + hydride) region, solid non-stoichiometric cerium hydride, $CeH_{2-\delta}$, will form (where δ is determined by the composition at the solidus line for the hydride). As further hydrogen is added, molten cerium at a higher temperature will solidify into hydride so that the solid-liquid interface will move toward hotter portions of the melt until completely solidified. In this manner, a single crystal of cerium hydride may be grown. The rate of growth is controlled merely by the rate of introducing hydrogen into the system.

The final crystal will, of course, have a concentration gradient along its length, since the solidus line shows a rapid increase in composition with temperature. However, the crystal can be homogenized by eliminating the thermal gradient and annealing under an appropriate hydrogen pressure. By diffusing additional hydrogen into the solid crystal, crystals having compositions anywhere between the solidus line and CeH_3 may be produced. Since the system must not pass through the region of two solid phases, $Ce(s) + CeH_{2-\delta}(s)$, on cooling to room temperature, the compositions must be above H/Ce = 1.8 before cooling.

Materials Preparation

Cerium metal* in the form of small rectangular billets approximately $\frac{1}{4} \times \frac{1}{4} \times 2$ in. is used in this preparation. The surfaces of the billets are filed down under a bath of mineral oil to ensure removal of any oxide layer. *Caution is advised as the metal may be pyrophoric.* The billets are rinsed in carbon tetrachloride to remove the mineral oil and are then transferred to the vacuum glove box. All subsequent handling of the metal is done in the vacuum glove box in a high-purity argon or helium gas environment.

*May be obtained from Ronson Metals Corporation, 45-65 Manufacturers Pl., Newark, N.J. 07105.

The hydrogen gas is purified by passing it through a hot Pd–Ag alloy in a hydrogen purifier.*

Apparatus

The single crystals are grown in a high-purity, 99.9%, tungsten metal boat having one end in a V-shaped configuration to minimize formation of multiple nucleation centers. The boats (approximately 4 in. long, $\frac{1}{2}$ in. wide, and $\frac{1}{4}$ in. deep) are preformed from 0.005-in. tungsten foil to the proper shape, after which the seams are joined *in vacuo* by electron-beam welding techniques. After having been welded, the boats are tested for leaks and cleaned thoroughly. They are then heated in hydrogen in a silica reaction tube at 1025°C. for 2 hours and cooled slowly to room temperature. The cleaned boats are transferred in the reaction tube to a vacuum glove box for storage until needed.

The crystal-growing apparatus consists of a gas-handling and vacuum system with calibrated volumes (see Fig. 24). A precision metering valve† (a), calibrated at 90 p.s.i.g. to determine flow rates, is located between the hydrogen purifier (b) and the reaction tube. A precision, compound, vacuum-pressure, dial gage (c) is used to monitor the gas pressure above the metering valve. The high-pressure side of the system is constructed of $\frac{1}{4}$-in. stainless-steel type-304 metal tubing with silver-brazed joints. The right-hand side of the system (see Fig. 24), beyond the metal-to-glass seal, is constructed of Pyrex glass up to the silica reaction tube.

The furnace is a Marshall high-temperature testing furnace‡ (d), with nine zones. Shunt resistances are placed selectively across the winding of the furnace to establish a thermal gradient across the 4-in. length of the boat, with the V-shaped end of the boat at the cool end of the gradient. A Pt–Pt–10%Rh thermocouple is used

*Milton Roy Company, St. Petersburg, Fla. 33700.

†Vacuum Accessories Corporation of America, Huntington Station, N.Y. 11746.

‡Varian Vacuum Division, (NRC), Sunnyvale, Calif. 94086.

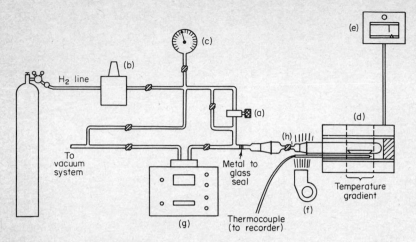

Fig. 24. *Apparatus for single-crystal* CeH_2 *preparation.* (a), *precision metering valve;* (b), *hydrogen purifier;* (c), *compound vacuum-pressure gage;* (d), *furnace;* (e), *temperature controller;* (f), *cooling fan;* (g), *precision pressure gage.*

to monitor the temperature, which is kept within ±1°C. with a temperature controller E. Any change of temperature within this range is by slow drift rather than by undesirable short fluctuations. A strip-chart recorder is used to record the temperature.

A cooling fan F is installed to keep the greased joints adjacent to the furnace cool. Hydrogen pressures are monitored with a Texas Instruments quartz Bourdon-tube pressure gage* G.

Procedure

■ *Note. All the procedures in this preparation require the use of anaerobic techniques to protect the cerium metal and subsequent compounds from oxidation.* The availability of a high-efficiency glove box and suitable transfer devices for weighing, transferring, and storing samples is essential. A glove box with a working environment of less than 1 p.p.m of water and oxygen in argon or helium gas is recommended.[5]

*Texas Instruments Incorporated, Houston, Tex. 77006.

A sample of cerium metal (approximately 15 g.), previously cleaned and weighed, is placed in a V-shaped tungsten boat. The boat is positioned inside a 26-mm.-o.d., heavy-wall, silica reaction tube, and a curved shield of tungsten foil is placed over the boat to prevent the possibility of silica flakes (formed by reaction with cerium vapor) from falling into the melt. The silica reaction tube is transferred anaerobically from the glove box and installed on the vacuum and gas-handling system.

The furnacc is positioned such that the desired temperature gradient corresponds exactly to the position of the boat. A narrow ($\frac{1}{4}$-in.), silica thermocouple tube is attached to the reaction tube along its length. The temperature is monitored normally at the V (cold end) of the boat. However, the gradient may be checked during the course of a run by moving the thermocouple inside the thermocouple tube. The shunt resistances on the furnace are set to give a linear gradient of about 20°C. across the length of the boat.

The system is evacuated completely, and the cerium metal is heated until the colder end of the boat reaches about 1020°C. (hotter end, *ca.* 1040°C.). The exact temperatures are not critical as long as the colder end of the boat is above 1010°C., the peritectic temperature. After the temperature is stabilized by the controller, purified hydrogen gas is bled into the system. The flow rate through the metering valve is set at about 20 cc./hour. Such a rate permits the solid-liquid interface to move at approximately 3 mm./hour for a 15-g. weight of sample and the thermal gradient employed. The rate of crystal growth decreases with the amount of cerium used and with the steepness of the temperature gradient at a given flow rate.

■ *Caution. Proper safety precautions are advised during all operations in the heating of both cerium and hydrogen. In the event of a leak, the sudden inflow of air reacting with hot cerium and hydrogen could lead to an explosion. Consequently, safety shields should be installed in the vicinity of the furnace and glass portions of the system.*

According to Fig. 23, the formation of cerium hydride begins when the hydrogen/metal ratio in the cerium melt exceeds 0.55 at 1020°C. The initial portion of crystal formed appears to have a hydrogen/metal ratio of 1.15 (that is, $\delta = 0.85$) from Lynds' data (Fig. 23). However, indications from more recent work are that the solidus line above the peritectic temperature should be shifted toward higher hydrogen contents which would give a composition at 1020°C. of $CeH_{1.33}$. The final portion of hydride crystal solidifying at 1040°C. has the composition of $CeH_{1.28}$ from Lynds' data, or $CeH_{1.5}$ from the present data.

After the melt is solidified completely, additional hydrogen is diffused into the crystal to bring the overall composition to H/Ce ≥ 1.8. The equilibrium hydrogen pressure at this composition and temperature is 600–700 torr. The cerium hydride crystal is cooled to 500°C. at a rate of about 30°C./hour. The thermal gradient is removed, and the crystal is annealed at 500°C. for about 24 hours to homogenize the composition. At this point additional hydrogen may be diffused into the crystal to bring it to any desired composition between H/Ce = 1.8 and 3. The crystal should be annealed for at least 24 hours after the hydrogen pressure becomes constant.

After annealing, the crystal is cooled slowly to room temperature under a minimum amount of hydrogen to avoid too much hydrogen uptake. Stopcock *H* is closed, and the reaction tube containing the solidified melt in the tungsten boat is transferred to the glove box. The crystal is removed carefully from the tungsten boat by peeling away the metal with mechanical tools, with extreme care used to avoid undue cracking of the crystal. The crystal tends to stick to the tungsten, but chemical analysis reveals no trace of tungsten in the crystal.

The melt consists of several crystals rather than one large crystal, due probably both to formation of several nucleation centers during growth, and to fracturing of the crystals on cooling and on removal from the boat. The average crystal size is 1 X 0.5 X 0.2 cm. The crystals are stored inside the glove box in Pyrex glass vials having ground-glass tops and sealed with caps having fluorocarbon seals.

Properties

Cerium hydride, single crystals are blue-black to blue-gray in color. The faces at the surface of the melt have a greenish tinge and a very high metallic luster. The crystals are quite brittle and fracture easily. It is difficult to cleave the crystals without some fracture also occurring. In cases where a sharply cleaved surface was obtained, it was always found to be the (111) face. A rather high degree of crystalline perfection is indicated by the Laue pattern, as seen in Fig. 25.

Some single crystals were taken out into the laboratory atmosphere to determine the degree of reactivity. They were

Fig. 25. Laue pattern of the (111) face of a CeH$_2$ *single crystal (*Mo *radiation).*

unusually stable and maintained a lustrous appearance for several days. Such behavior is in contrast to that of polycrystalline cerium hydride, which reacts readily with air and is frequently pyrophoric at the higher hydrogen contents. When cleaved or scratched with a pointed tool in air, however, the crystals gave off sparks, and one batch of crystals ignited spontaneously on exposure to humid air. Apparently, a thin protective oxide layer which serves to inhibit further reaction forms on exposure to air and moisture.

Electrical resistivity measurements[6] on a variety of compositions $(CeH_{1.9}\text{-}CeH_{2.85})$ in the temperature range of -130 to $25°C$. gave values ranging from 10^{-4} to 500 Ω-cm., depending upon composition and temperature.

References

1. G. G. Libowitz, "The Solid State Chemistry of Binary Metal Hydrides," W. A. Benjamin, Inc., New York (1965).
2. B. Stalinski, *Bull. Acad. Polon. Sci., Classe III,* **7**, 269 (1959); R. C. Heckman, *J. Chem. Phys.,* **40**, 2958 (1964); *ibid.* **46**, 2158 (1967).
3. T. C. Harman, J. I. Genco, W. P. Allred, and H. L. Goering, *J. Electrochem. Soc.,* **105**, 731 (1958).
4. E. P. Stambaugh, J. F. Miller, and R. C. Himes, Metallurgy of Elemental and Compound Semiconductors, "Metallurgical Society Conference Number 12," R. O. Grubel (ed.), Interscience Publishers, Inc., New York, 1961.
5. J. G. Pack and G. G. Libowitz, *Rev. Sci. Instr.,* **40**, 414 (1969).
6. G. G. Libowitz and J. G. Pack, in "Crystal Growth," H. S. Peiser (ed.), pp. 129–132, Pergamon Press, New York, 1967.
7. G. G. Libowitz and J. G. Pack, *J. Chem. Phys.,* **50**, 3557 (1969).
8. L. Lynds, *Bull Am. Phys. Soc.,* **8**, 473 (1963).
9. R. N. R. Mulford and C. E. Holley, Jr., *J. Phys. Chem.,* **59**, 1222 (1955).
10. R. Streck and K. Dialer, *Z. Anorg. Allgem. Chem.,* **306**, 141 (1960).

INDEXES

INDEX OF CONTRIBUTORS

SUBJECT INDEX

Names used in this cumulative Subject Index for Volumes XI through XIV as well as in the text, are based for the most part upon the "Definitive Rules for Nomenclature of Inorganic Chemistry," 1957 Report of the Commission on the Nomenclature of Inorganic Chemistry of the International Union of Pure and Applied Chemistry, Butterworths Scientific Publications, London, 1959; American version, *J. Am. Chem. Soc.*, 82, 5523–5544 (1960); and the latest revisions (in press as a Second Edition (1970) of the Definitive Rules for Nomenclature of Inorganic Chemistry); also on the Tentative Rules of Organic Chemistry—Section D; and "The Nomenclature of Boron Compounds" [Committee on Inorganic Nomenclature, Division of Inorganic Chemistry, American Chemical Society, published in *Inorganic Chemistry, 7,* 1945 (1968) as tentative rules following approval by the Council of the ACS]. All of these rules have been approved by the ACS Committee on Nomenclature. Conformity with approved organic usage is also one of the aims of the nomenclature used here.

In line, to some extent, with *Chemical Abstracts* practice, more or less inverted forms are used for many entries, with the substituents or ligands given in alphabetical order (even though they may not be in the text); for example, derivatives of arsine, phosphine, silane, germane, and the like; organic compounds; metal alkyls, aryls, 1,3-diketone and other derivatives and relatively simple specific coordination complexes: *Iron, cyclopentadienyl-* (also at *Ferrocene*); *Cobalt(II), bis(2,4-pentanedionato)-* [instead of *Cobalt(II) acetylacetonate*]. In this way, or by the use of formulas, many entries beginning with numerical prefixes are avoided; thus *Vanadate(III), tetrachloro-.* Numerical and some other prefixes are also avoided by restricting entries to group headings where possible: *Sulfur imides*, with the formulas; *Molybdenum carbonyl*, $Mo(CO)_6$; both *Perxenate*, $HXeO_6^{3-}$, and *Xenate(VIII)*, $HXeO_6^{3-}$. In cases where the cation (or anion) is of little or no significance in comparison with the emphasis given to the anion (or cation), one ion has been omitted; e.g., also with less well-known complex anions (or cations): $CsB_{10}H_{12}CH$ is entered only as *Carbaundecaborate(1−)*, *tridecahydro-* (and as $B_{10}CH_{13}^-$ in the Formula Index).

Under general headings such as *Cobalt(III) complexes* and *Ammines*, used for grouping coordination complexes of similar types having names considered unsuitable for individual headings, formulas or names of specific compounds are not usually given. Hence it is imperative to consult the Formula Index for entries for specific complexes.

Two entries are made for compounds having two cations and for addition compounds of two components, with extra entries or cross references for synonyms. Unsatisfactory or special trivial names that have been retained for want of better ones or as synonyms are placed in quotation marks.

Boldface type is used to indicate individual preparations described in detail, whether for numbered syntheses or for intermediate products (in the latter case, usually without stating the purpose of the preparation). Group headings, as *Xenon fluorides*, are in lightface type unless all the formulas under them are boldfaced.

As in *Chemical Abstracts* indexes, headings that are phrases are alphabetized straight through, letter by letter, not word by word, whereas inverted headings are alphabetized first as far as the comma and then by the inverted part of the name. Stock Roman numerals and Ewens-Bassett Arabic numbers with charges are ignored in alphabetizing unless two or more names are otherwise the same. Footnotes are indicated by n. following the page number.

Rhenium(III), tetrakis(μ-p-anisato)
dichlorodi-, with rhenium-
rhenium quadruple bonds,
13:86
———, tetrakis(μ-benzoato)di-
bromodi-, with rhenium-
rhenium quadruple bonds,
13:86
———, tetrakis(μ-benzoato)di-
chlorodi-, with rhenium-
rhenium quadruple bonds,
13:86
———, tetrakis[μ-(p-bromobenz-
oato)] dichlorodi-, with
rhenium-rhenium quadruple
bonds, 13:86
Rhenium(V) chloride, decomposition to Re_3Cl_9, 12:193
Rhenium compounds with quadruple bonds, 13:81–86
Rhenium(IV) oxide, β-, single crystals, 13:142
Rhodium(I), chlorobis(cyclooctene)-, 14:93
———, hydridotetrakis(triphenyl
phosphite)-, 13:109
Rhodium(II), tetrakis(μ-acetato)di-,
13:90
Rhodium(III), pentaamminechloro-,
dichloride, 13:213
———, pentaamminehydrido-, sulfate, 13:214
Rhodium(I) complexes, with triaryl
phosphites, 13:105
with triphenylarsine,
11:100
with triphenylphosphine and
CH_3 and F derivatives,
11:99, 190
Rhodium(III) complexes, cation
with ethylenediamine (tris),
resolution of, 12:269, 272
Rubidium aquapentachloromolyb-
date(III), 13:171

Rubidium hexachloromolybdate-
(III), 13:172
Ruthenium(0), dodecacarbonyl-
tri-, *triangulo*-, 13:92
Ruthenium(II), chlorohydridotris-
(triphenylphosphine)-, 13:131
———, hexaammine-, dichloride,
13:208n., 209
———, hexaammine-, tetrachloro-
zincate, 13:210
Ruthenium(III), hexaammine-,
tribromide, 13:211
———, pentaamminechloro-,
dichloride, 13:210
Ruthenium(II) complexes, ammines
and (N_2)-containing, 12:2, 3, 5
with azide, (N_2), and ethylenediamine, 12:23
with triphenylphosphine,
12:237–240
Ruthenium(III) complexes,
ammine, 12:3–4, 7
Ruthenium(IV) complexes, anion,
μ-oxo-bis[pentachlororuth-
enate(IV)], 11:70
Ruthenium(IV) oxide, single crystals, 13:137

Safety, in ammonium cyanate preparation, 13:18
in arsine preparation, 13:14
in carbon monoxide use, 13:128
in iridium and ruthenium dioxide preparation, 13:137
in iridium carbonyl preparation,
13:96
in molybdenum pentafluoride
preparation, 13:146
in osmium dioxide preparation,
13:140
in perbromate preparation, 13:1
in phosphine silyl derivative
preparation, 13:27

FORMULA INDEX

The Formula Index, as well as the Subject Index, is a cumulative index for Volumes XI through XIV. The chief aim of this index, like that of other formula indexes, is to help in locating specific compounds or ions, or even groups of compounds, that might not be easily found in the Subject Index, or in the case of many coordination complexes are to be found only as general entries in the Subject Index. *All* specific compounds, or in some cases ions, with definite formulas (or even a few less definite) are entered in this index or noted under a related compound, whether entered specifically in the Subject Index or not. As in the latter index, **boldface type** is used for formulas of compounds or ions whose preparations are described in detail, in at least one of the references cited for a given formula.

Wherever it seemed best, formulas have been entered in their usual form (*i.e.*, as used in the text) for easy recognition: Si_2H_6, XeO_3, NOBr. However, for the less simple compounds, including coordination complexes, the significant or central atom has been placed first in the formula in order to throw together as many related compounds as possible. This procedure often involves placing the cation last as being of relatively minor interest (*e.g.*, alkali and alkaline earth metals), or dropping it altogether: MnO_4Ba; $Mo(CN)_8K \cdot 2H_2O$; $Co(C_5H_7O_2)_3Na$; $B_{12}H_{12}{}^{2-}$. Where there may be almost equal interest in two or more parts of a formula, two or more entries have been made: Fe_2O_4Ni and $NiFe_2O_4$; $NH(SO_2F)_2$, $(SO_2F)_2NH$, and $(FSO_2)_2NH$ (halogens other than fluorine are entered only under the other elements or groups in most cases); $(B_{10}CH_{11})_2Ni^{2-}$ and $Ni(B_{10}CH_{11})_2{}^{2-}$.

Formulas for organic compounds are structural or semistructural so far as feasible: $CH_3COCH(NHCH_3)CH_3$. Consideration has been given to probable interest for inorganic chemists, *i.e.*, any element other than carbon, hydrogen, or oxygen in an organic molecule is given priority in the formula if only one entry is made, or equal rating if more than one entry: only $Co(C_5H_7O_2)_2$, but $AsO(+)-C_4H_4O_6Na$ and $(+)-C_4H_4O_6AsONa$. Names are given only where

223

the formula for an organic compound, ligand, or radical may not be self-evident, but not for frequently occurring relatively simple ones like C_5H_5 (cyclopentadienyl), $C_5H_7O_2$ (2,4-pentanedionato), C_6H_{11} (cyclohexyl), C_5H_5N (pyridine). A few abbreviations for ligands used in the text are retained here for simplicity and are alphabetized as such: "en" (under "e") stands for ethylenediamine, "py" for pyridine, "bipy" for bipyridine, "pn" for 1,2-propanediamine (propylenediamine), "fod" for 1,1,1,2,2,3,3-hepta-fluoro-7,7-dimethyl-4,6-octanedionato, "thd" for 2,2,6,6-tetramethylhep-tane-3,5-dionato, "dien" for diethylenetriamine or N-(2-aminoethyl)-1,2-ethanediamine, "chxn" for trans-1,2-cyclohexanediamine, "DH" for di-methylglyoximato and "D" for the dianion, $(CH_3)_2C_2N_2O_2{}^{2-}$.

The formulas are listed alphabetically by atoms or by groups (considered as units) and then according to the number of each in turn in the formula rather than by total number of atoms of each element. This system results in arrangements such as the following:

NHS_7

$(NH)_2S_6$ (instead of $N_2H_2S_6$)

$NH_3B_{10}CH_{12}$

$[Mo(CO)_3C_5H_5]K$

$[Mo(CO)_3C_5H_4CH_3]$

FNO

$(FSO_2)_2NH$ (instead of $F_2S_2O_4NH$)

FSO_3H

F_2SO_3

$[Cr(en)_3][Ni(CN)_5]$ ["en" instead of $(NH_2)_2C_2H_4$ or $N_2H_4C_2H_4$]

$[Cr(NH_3)_6][Ni(CN)_5]$

Footnotes are indicated by n. following the page number.

Ag[BF_4], 13:57n.

$AgFeO_2$, 14:139

$Ag(NO_2)_2$, 13:205

$Al(C_2H_5)_3$, 13:126

$AlH_3 \cdot 0.3[(C_2H_5)_2O]$, 14:47

AlH_4Li, 11:173

$[As(CH_3)_2]C_3H_6[P(CH_3)_2]$, 14:20

$[\{As(C_6H_5)_3\}PtCl(C_8H_{12})][BF_4]$
Chloro(1,5-cyclooctadiene) (triphenylarsine)platinum(II) tetrafluoroborate, 13:64

$[\{As(C_6H_5)_3\}PtCl(C_{13}H_{19}O_2)]$
[8-(1-Acetylacetonyl)-4-cyclo-octen-1-yl] chloro(triphenyl-arsine)platinum(II), 13:63

$[\{As(C_6H_5)_3\}_2Pd(SCN)_2]$, 12:221; (NCS)$_2$ isomer, 12:221

$[\{As(C_6H_5)_3\}_2Rh(CO)Cl]$, trans-, 11:100

$[As(C_6H_5)_4]ThI_6$, 12:229

$[As(C_6H_5)_4][VBr_4]$, 13:168

$[As(C_6H_5)_4][VCl_4]$, 13:165

$[As(C_6H_5)_4]_2UI_6$, 12:230

AsD_3, 13:14

$(AsF_6)O_2$, 14:39

AsH_3, 13:14

$AsNa$, 13:15

$As(OCH_3)_3$, 11:182

$As(OC_2H_5)_3$, 11:183

$AsO(+)C_4H_4O_6Na$, 12:267

$As(OC_4H_9)_3$, 11:183

$AuBrTe_2$, 14:170

$AuClTe_2$, 14:170

$AuITe$, 14:170

$AuITe_2$, 14:170

BBrH$_2$·N(CH$_3$)$_3$, 12:118

BBr$_2$H·N(CH$_3$)$_3$, 12:123

BBr$_3$, 12:146; compound with N(CH$_3$)$_3$, 12:141, 142

[B(4-CH$_3$C$_5$H$_4$N)$\{$N(CH$_3$)$_3\}$H$_2$]I, 12:132; PF$_6^-$salt, 12:134

[B(4-CH$_3$C$_5$H$_4$N)$_4$]Br$_3$, 12:141; PF$_6^-$salt, 12:143

[B(C$_3$H$_3$N$_2$)$_2$H$_2$]K Potassium dihydrobis(1-pyrazolyl)borate, 12:100

[B(C$_3$H$_3$N$_2$)$_2$H$_2$]$_2$Ni, 12:104

[B(C$_3$H$_3$N$_2$)$_3$H]K, 12:102

[B(C$_3$H$_3$N$_2$)$_3$H]$_2$Co, 12:105

[B(C$_3$H$_3$N$_2$)$_4$]K, 12:103

[B(C$_3$H$_3$N$_2$)$_4$]$_2$Mn, 12:106

[B(C$_5$H$_5$N)$_3$]HBr$_2$, 12:139; PF$_6^-$salt, 12:140

[B(C$_5$H$_7$O$_2$)$_2$][HCl$_2$], 12:128

[B(C$_5$H$_7$O$_2$)$_2$][SbCl$_6$], 12:130

B(C$_6$H$_5$)$_3$, 14:52

B(C$_6$H$_5$)$_4$[N(CH$_3$)$_3$H], 14:52

BCl(C$_6$H$_5$)$_2$, 13:36

BClH$_2$·N(CH$_3$)$_3$, 12:117

BCl$_2$(C$_6$H$_5$), 13:35

BCl$_3$, adduct with CH$_3$CN, 13:42

(BF$_2$D)$_2$Co(CH$_3$)·H$_2$O Complex from methylaquocobaloxime and BF$_3$, 11:68

[BF$_4$]Ag, 13:57n.

BH$_3$ Compounds with ethylenediamine, P(C$_6$H$_5$)$_3$, P(C$_2$H$_5$)$_3$, N(C$_2$H$_5$)$_3$, morpholine, and tetrahydrofuran, 12:109–115; with P(CH$_3$)$_3$, 12:135, 136

BH$_4^-$, 11:28

BH$_4$Cs·B$_{12}$H$_{12}$Cs$_2$, 11:30

BIH$_2$·N(CH$_3$)$_3$, 12:120

[B(NH$_2$)(O$_2$CCH$_3$)$_2$]$_2$, 14:55

BO$_3$(CH$_3$)$_3$, 12:50, 52n.

[B$\{$P(CH$_3$)$_3\}_2$H$_2$]I, 12:135

B$_2$(C$_3$H$_3$N$_2$)$_2$H$_4$ Pyrazabole, 12:107; substituted, 12:99, 108

B$_2$H$_6$, 11:15

B$_2$N$_2$H$_4$(O$_2$CCH$_3$)$_4$, 14:55

B$_3$Cl$_3$N$_3$H$_3$, 13:41

B$_3$Cl$_3$N$_3$(CH$_3$)$_3$, 13:43

B$_3$H$_8^-$, 11:27

B$_8$H$_8^{2-}$, 11:24

B$_9$H$_9^{2-}$, 11:24

[B$_{10}$CH$_{10}$N(CH$_3$)$_2$]$_2$Ni$_2^-$, 11:45

[B$_{10}$CH$_{10}$N(CH$_3$)$_2$H]$_2$Ni, 11:45

(B$_{10}$CH$_{10}$NH$_2$)$_2$Ni^{2-}, 11:44

B$_{10}$CH$_{10}$NH$_3^{2-}$, 11:41

(B$_{10}$CH$_{10}$NH$_3$)$_2$Ni, 11:43; analogous Fe(III) and Co(III) complexes, 11:42

(B$_{10}$CH$_{10}$OH)$_2$Ni^{2-}, 11:44

B$_{10}$CH$_{11}^{3-}$11:40n., 41

(B$_{10}$CH$_{11}$)$_2$Ni^{2-}, 11:42; analogous Fe(III) and Co(III) complexes, 11:42

B$_{10}$CH$_{12}$N(CH$_3$)$_2$(C$_3$H$_7$), 11:37

B$_{10}$CH$_{12}$N(CH$_3$)$_3$, 11:35

B$_{10}$CH$_{12}$N(C$_3$H$_7$)H$_2$, 11:36

B$_{10}$CH$_{12}$NH$_3$, 11:33

B$_{10}$CH$_{13}^-$, 11:39

B$_{10}$C$_2$H$_{10}$(CH$_2$O$_2$CCH$_3$)$_2$, 11:20

B$_{10}$C$_2$H$_{12}$, 11:19

B$_{10}$H$_{10}^{2-}$, 11:28, 30

B$_{10}$H$_{14}$, 11:20n., 34n.

B$_{11}$H$_{11}^{2-}$, 11:24

B$_{11}$H$_{13}^{2-}$, 11:25

B$_{11}$H$_{14}^-$, 11:26

B$_{12}$H$_{12}^{2-}$, 11:28, 30

B$_{12}$H$_{12}$Cs$_2$·BH$_4$Cs, 11:30

B$_{12}$H$_{12}$Cs$_2$·ClCs, 11:30

BaTiO$_3$, 14:142

BiBrS, 14:172

BiIS, 14:172

Bi$_4$Ti$_3$O$_{12}$, 14:144

BrF$_3$, 12:232

BrO$_3$F, 14:30

BrO$_4$H, 13:1

BrO$_4$K, 13:1

C(C$_6$H$_5$)$_3$H, 11:115

CF$_2$(OF)$_2$, 11:143